# Transitions and Borders between Animals, Humans and Machines 1600-1800

*Edited by*

Tobias Cheung

BRILL

LEIDEN • BOSTON
2010

*Cover illustration*: Malpighi, Marcello: *Opera omnia*. 2 vols. London: Apud Robertum Scott & Georgium Wells 1686. Vol. 1, plate XXXIX, figure 141. Web of ligneous fibres of a root.

Originally published as Volume XV, Nos. 1-2 (2010) of Brill's journal *Early Science and Medicine*.

This book is printed on acid-free paper.

Library of Congress Cataloging-in-Publication Data

Transitions and borders between animals, humans, and machines, 1600-1800 / edited by Tobias Cheung.
p. cm.
English and German.
"Originally published as volume XV, nos. 1-2 (2010) of Brill's journal, Early science and medicine"—T.p. verso.
Includes index.
ISBN 978-90-04-19181-5 (hardback : alk. paper)
1. Philosophical anthropology—History. 2. Mechanism (Philosophy)—History. 3. Animals (Philosophy)—History. 4. Philosophy, Modern. I. Cheung, Tobias.

BD450.T648 2010
128.09—dc22

2010039951

ISBN 978 90 04 19181 5

# Contents

# Transitions and Borders between Animals, Humans and Machines 1600-1800: Introduction

Tobias Cheung

The search for a new foundation of the order of things, that characterizes the period between Descartes and Kant, is closely related to three questions: What is an animal? What is a human? What is a machine? The various answers that have been given to the questions occur in a field of dynamic interactions between theories of knowledge and of matter, experiments, observations, moral, theological and scientific claims, analogies, metaphors, imitations, and specific objects or artifacts.[1]

The main objective of this volume is to shed new light on different disciplinary, methodological and conceptual perspectives of this field as well as on contrasting strategies from 1600 to 1800. These perspectives and strategies are not just interpretations of a kind of ahistorically conceived typology of 'the animal', 'the human', and 'the machine'. Rather, they define the mode of existence of animals, humans and machines within specific historical constellations and within manifold worlds of things. These worlds include monsters,

---

1) For further references and discussions, see James J. Sheehan & Morton Sosna, eds., *The Boundaries of Humanity: Humans, Animals, Machines* (Berkeley, 1991); Erica Fudge, Susan Wiseman & Ruth Gilbert, eds., *At the Borders of the Human: Beasts, Bodies and Natural Philosophy in the Early Modern Period* (Basingstoke, 2002); Jessica Wolfe, *Humanism, Machinery, and Renaissance Literature* (Cambridge, 2004); Thierry Gontier, ed., *Animal et animalité dans la philosophie de la Renaissance et de l'Âge Classique* (Louvain-la-Neuve, 2005); Charles Wolfe, ed., *Monsters and Philosophy* (London, 2005); and Frank Palmeri, *Humans and Other Animals in Eighteenth-Century British Culture: Representation, Hybridity, Ethics* (Burlington, 2006).

angels, spirits, plants, polyps, fibres, germs, and the seemingly non-living. The historical constellations, the concrete experimental investigations and the manifold worlds of things thus stabilize and destabilize the triangular field of transitions and borders between animals, humans, and machines.

The five essays of this volume reconstruct the shifting ground of this triangular field. Ann Thomson's essay relates the problem of the human-animal-distinction in England at the turn of the eighteenth century to debates on different types of souls and matter theories that have been used to explain self-movement, sensation, instinct-like behavior and reflexivity. Comparing various theological positions, she discusses the dilemma that a materialistic conception of an animal soul, that can sense and move itself, but which is not immortal like the human soul, could also appear to imply the possibility of thinking matter. Charles Wolfe's essay on living minima and endowed molecules in the matter theories of Maupertuis and Diderot transforms this dilemma into the framework of French Materialism in the second half of the eighteenth century. When the strategy of the endowed molecule comes into play, as Wolfe puts it, then the attribution of a self-organizing activity to the molecule, in which "the materials are the workers themselves," has to be discussed. My own contribution focuses on a different and yet closely related strategy in the French context: the fibre œconomies of organized bodies of Charles Bonnet and Diderot. In these œconomies, stimuli-reaction-schemes and vibrations connect inside-outside-events and transform sensations into thinking or "physical ideas." The agents of these activities, or the material workers, are the fibres themselves. In her essay, Yvonne Wübben switches the perspective from stimuli-reaction-schemes of fibre œconomies to the historical context of the notion of reflex and its experimentalization, especially in the work of Thomas Willis. While fibre œconomies are related to technical metaphors of looms, Hanns-Peter Neumann shows in a longue-durée study that models of regulation and self-adjustment, which already Cusanus and Ficino used to characterize both mental and organic processes, are often based on the order of a different machine: the clock.

# Animals, Humans, Machines and Thinking Matter, 1690-1707

**Ann Thomson**
*Université Paris 8**

**Abstract**
This article looks at the debate on the soul in England at the turn of the eighteenth century and at the role played within it by the question of animal soul, which had both theological and scientific ramifications. It discusses the difficulty of accounting for animal behaviour without either adopting the animal-machine hypothesis or according animals an immaterial and hence immortal soul. While those who denied the existence of an immaterial human soul and refused any fundamental distinction between humans and other animals were accused of reducing humans to machines, this article shows that the issues were in fact more complex. The fundamental question was that of the nature of matter; the main danger for many theologians seemed to lie in the attribution of innate life and sensibility to matter, which opened the door to materialism and undermined Christian doctrine.

**Keywords**
animal, soul, matter, machine, reason, instinct, Henry Layton, William Coward, Anthony Collins.

> Methinks I hear the *Psychomuthist* arguing, that according to my notion: man is a meer piece of mechanism, a curious frame of clock-work, and only a reasoning engine; to which I answer [...] if man be a piece of mechanism, 'tis no more than what the scripture says of him in other cases comparatively; but I must remind my adversary that man is such a curious piece of mechanism, as shews only an almighty power could be the first & sole artificer, *viz.* to make a reasoning engine

* DEPA, Université Paris 8, 2 rue de la Liberté, 93526 Saint-Denis Cedex, France (ann.thomson@wanadoo.fr).

> out of dead matter, a lump of insensible earth to live, to be able to discourse; to pry and search into the very nature of heaven & earth.[1]

The claim that "man" is a "machine" is normally associated with the scandalous figure of Julien Offray de La Mettrie, whose provocatively titled *Homme machine* has resonated ever since its publication in 1747. However, as we can see from the above quotation, the debate was far from new, as similar claims had been made earlier, albeit in a very different context and with different implications, and without La Mettrie's explicitly irreligious thrust. The present article explores how an important discussion on the existence of the human soul in England in the late seventeenth and early eighteenth centuries throws light on the implications of thinking about the links between machines, humans and animals. The issue at stake in this debate was the existence of an immaterial soul, held to be a precondition for its immortality, as reaffirmed at the Fifth Lateran Council in 1513. Attempts to demonstrate rationally the soul's immortality based on its necessary immateriality had exercised natural philosophers ever since that date. A strand within Protestant thinking denied its natural immortality, claiming that according to the Scriptures, the whole individual dies, to be resuscitated at the Last Judgement.[2] Those adopting this latter position, generally called Psychopannychists, often claimed that the soul died with the body, but by the late seventeenth century certain writers claimed that there was no separate immaterial soul and that "the whole man thinks" by means either of a material soul or of a particular organisation of matter in the brain. I have studied this debate in detail elsewhere.[3] Here I would like to look more closely at the role played by the reference to animals, the blurring of the frontier between humans and the other animals in these discussions, and the accu-

[1] William Coward, *Second Thoughts on Human Soul* (London, 1702), 122-124. "Psychomuthist" is Coward's word for the defenders of the soul's immateriality, possibly an echo of Henry More's "Psychopyrist".

[2] See Norman T. Burns, *Christian Mortalism from Tyndale to Milton* (Cambridge, Mass., 1972).

[3] Ann Thomson, *Bodies of Thought. Science, Religion, and the Soul in the Early Enlightenment* (Oxford, 2008).

sation that any assimilation of humans to animals implied reducing humans to machines. The debate could not take place in purely scientific terms and the theological aspects must be taken into account if one wishes to understand the discussion, for the question of the relationship between humans and animals involved discussion of the soul, which was a crucial theological issue and the subject of wide debate in the period. These issues were endlessly rehearsed in the second half of the seventeenth century, to a large extent in the wake of Cartesian philosophy and the substance dualism that it entailed, as the mechanistic philosophy in its various forms was perceived to contain the danger of promoting atheism. A widespread interpretation of what followed is that the Cartesian conception of animals as unthinking machines was extended to humans, resulting in the works of La Mettrie. This was indeed the danger immediately denounced by Descartes's critics.[4] However, much work on La Mettrie and other eighteenth-century writers has contributed to a more nuanced and complex understanding of what happened. For it was not only the mechanical philosophy that could be used to undermine orthodoxy, and certain medical theories contained within them the seeds of heterodox arguments concerning humans. John Henry, who has discussed the attention paid to medical theories of thinking matter by More and Cudworth, considers that these writings were generally ignored and the question was only brought centre-stage by Locke.[5] However crucial Locke's hypothesis may have been, there are nevertheless some indications that the medical writings were less marginal than Henry suggests. As we shall see in what follows, the heated debate on the human soul at the end of the seventeenth century did draw on medical writings, which helped to remove it from its confinement in the sphere of purely theological

4) See Thomas M. Lennon, *The Battle of the Gods and Giants* (Princeton, 1993), 318.

5) As John Henry shows, theological arguments against atheism in the second half of the seventeenth century tended to target mechanical philosophy, but Cudworth and More denounced the danger lying in medical views of living matter. John Henry, "The Matter of Souls: Medical Theory and Theology in Seventeenth-century England", in *The Medical Revolution of the Seventeenth Century*, ed. R. French and A. Wear (Cambridge, 1989), 87-113.

debate and bring it onto centre stage. Medical work, in particular concerning animals, played a role in these discussions.

The debate on the animal soul that followed in the wake of Cartesian philosophy[6] has been studied particularly for France.[7] But the implications of the question in England have attracted less attention, perhaps because English discussion of animals tended to take place in the context of debates about the human soul and because there was no specific polemic on the subject as there was in France. Nevertheless, Keith Thomas considers that "the idea of animal immortality seems to have made more headway in England than anywhere else in this period [...] buttressed by arguments from scripture and by observation of the mental capacities of the animal in question."[8] What I wish to investigate here is the extent to which the reflection on human nature in the late seventeenth century involved a discussion of the relationship between humans and other animals and how important were accusations of reducing humans to machines. According to John Yolton, reactions to two of the strands of materialism he identifies in Britain (Locke's suggestion about thinking matter and Collins's claim that organised matter can have properties that the individual parts of matter do not have) "frequently involved the fear that man would be viewed as a machine, as a piece of clockwork. Closely related was the debate over animals: do they reason or are they machines?"[9] This gives some indication of how these questions were linked and the parameters of the problem, while not exhausting the possibilities rehearsed in this period or the types of fear aroused, as we shall see.

---

6) According to Ralph Cudworth, it "was hardly ever called into doubt or question by any, before Cartesius, whether the souls of brutes had any sense, cogitation or consciousness in them or no", *The True Intellectual System of the Universe* (London, 1678, repr. Hildesheim, 1977), 39.

7) Leonora Cohen Rosenfield, *From Beast-Machine to Man-Machine. Animal Soul in French Letters from Descartes to La Mettrie* (New York, 1941); see also Heikki Kirkinen, *Les origines de la conception moderne de l'homme-machine. Le problème de l'âme en France à la fin du règne de Louis XV (1670-1715)* (Helsinki, 1960).

8) Keith Thomas, *Man and the Natural World* (London, 1983), 140.

9) John Yolton, *Thinking Matter* (Oxford, 1984), 30.

I shall discuss in what follows the arguments about what differentiates humans from animals with particular reference to the polemic on human nature and the immaterial soul at the turn of the eighteenth century centred around the writings of Henry Layton and Dr William Coward, quoted above. The books published by these authors denied the existence of an immaterial soul in humans, which could distinguish them from animals. The works of Coward in particular provoked a huge outcry, a large number of refutations (one of which was firmly entitled "Man is not a Machine"[10], sermons, articles, an attempt at censure by the Convocation of the Church of England and condemnation of two of Coward's books by the House of Commons. The controversy, to which John Toland contributed with his *Letters to Serena* (1704), was continued from a different perspective by Henry Dodwell, whose *Epistolary Discourse*, published in 1706, drew a reply from Samuel Clarke, sparking off an exchange of pamphlets between Clarke and Anthony Collins. I have elsewhere analysed the political circumstances which made the question particularly delicate in this period.[11] The political situation in the years following the Glorious Revolution was still tense, with the fear of an attempt by the High Church and Tories to bring back certain Catholic doctrines and to overturn the toleration, however limited, enshrined in 1689. The doctrine of an immortal soul became an element in the theologico-political confrontation during these years; Christians like Layton and Coward believed that the doctrine was a Platonic import into Christian teaching and wished to return to a pure religion purged of popish corruptions and free from "priestcraft". They also believed that to deny that matter could ever think was to limit God's omnipotence, which helps to explain why their main target was the "Cambridge Platonist" Henry More. Their works drew on contemporary philosophico-medical discussion, citing Thomas Hobbes, William Harvey, Kenelm Digby, Thomas Willis, Francis Glisson, Walter Charleton or Daniel Sennert.

---

10) William Nichols, *A Conference with a Theist*, part V, 3rd ed, 2 vols. (London, 1723), 2: 213ff. See below.

11) Thomson, *Bodies of Thought*, 29-44.

The present article will evoke this background in order to bring out the interlinking scientific and theological discussion around the question of human nature and the relationship between humans and animals, linked to conceptions of matter; we shall also look at how far the fear of turning humans into machines was an important factor.[12] The debate had of course been complicated by the dominance of the Cartesian paradigm, opposed to Aristotelianism, which enshrined the view of matter as passive extension and the necessary immateriality of thinking substance. Cartesian philosophy had been popularised in England mainly thanks to Henry More, but the problems it entailed continued to nag and led More to express greater reticence towards Cartesianism and to develop views with certain somewhat heterodox implications in his own writings.[13] Another English defender of the Cartesian position was the Catholic Sir Kenelm Digby, who was a member of the Mersenne circle when in exile in France in the 1640s and who developed a form of what has been called "Aristotelian atomism".[14] In the first of his *Two Treatises* (1644) intended to demonstrate the immortality of the soul, he compared plants to a waterworks he saw in Toledo and animals to the ingenious machine at the mint in Segovia, which required outside direction. He attempted to show how the various stories told to prove animals' ability to reason could all be explained by mechanism,[15] refusing explanations in terms of "secret instincts, and antipathies, and sympathies, whereof we can give no account."[16] Digby was far from being the only one who dismissed "instinct" as an occult quality, although such a position was more often taken

12) I shall not go into the question of free will and determinism, but it should be remembered that an important implication of the notion of machine was the lack of any free will.

13) These have been explored by John Henry: "A Cambridge Platonist's Materialism. Henry More and the Concept of Soul", *Journal of the Warburg and Courtauld Institutes*, 39 (1986), 172-195.

14) John Henry, "Atomism and Eschatology: Catholic and Natural Philosophy in the Interregnum", *British Journal for the History of Science*, 15 (1982), 211-239.

15) Kenelm Digby, *Two Treatises* (Paris, 1644), 205-8, 306-339.

16) *Two Treatises*, 332.

up by those who defended a type of reason in animals.[17] However, the word was also used to refer to precisely the natural or mechanical disposition that Digby was defending and was in fact employed by Descartes to mean a natural impulse towards self-preservation or pleasure, which exists in humans and animals.[18] It appealed to defenders of animal mechanism as it enabled actions which were frequently quoted as evidence of a form of reasoning to be given a purely mechanical explanation in terms of a natural disposition which did not require thought; the regularity of animal behaviour was cited as proof that it was not the result of reflection on their part but of nature. Instinct was thus distinguished from reason.[19]

But here too the notion was open to diverse interpretations on the part of those for whom purely mechanical explanations were insufficient or dangerous: for Henry More—who defined natural instinct as an "instigation or impetus" in animals to act "without reason, counsel, deliberation, or acquired knowledge" for the common good of the species—it was caused by the Spirit of Nature,[20] while for Ralph Cudworth animals' natural instincts demonstrated the workings of plastick nature.[21] Thomas Willis, who accorded animals knowledge and an inferior type of reason, took a somewhat different view of natural instinct. He defined it in all living creatures as "an innate conservation of themselves", explaining that animals are provided by the creator with an innate knowledge of what is good and what is harmful for them, "by which they are instigated though a secret impulse and blind power to propagation of their kind." He distinguished this innate knowledge, which he also called a "law or rule fixed in their hearts", from "acquired cogni-

---

[17] For example, Marin Cureau de La Chambre, *Traité de la connoissance des bêtes* (1645). See also below.

[18] Letter to Mersenne, 16 October 1639, quoted by Geneviève Rodis-Lewis, *L'anthropologie cartésienne* (Paris, 1990), 163.

[19] For more examples, see Werner Krauss, "Zur Seelentheorie im 18. Jahrhundert", in *Zur Anthropologie des 18. Jahrhunderts* (Berlin, 1978), 136-175.

[20] Henry More, *The Immortality of the Soul*, ed. Alexander Jacob (Dordrecht, 1987), 267.

[21] *The True Intellectual System*, 158: see François Duchesneau, *Les modèles du vivant de Descartes à Leibniz* (Paris, 1998), 175.

tion" which perfects it.[22] Instinct was also invoked by physico-theologians, but while John Ray recounted many "strange instincts of animals" to demonstrate God's wisdom in guiding them,[23] an anonymous author who was making the same point concerning God's wisdom rejected the term "instinct", calling those who appealed to it "ignorant people". Like Digby he denounced its obscurity: "let these persons explain themselves: what do they understand by this word *instinct?* All the world makes use of it, and there is scarce one man that has any *idea* of it." He was objecting to those who made it "the motion of a blind nature devoid of any knowledge", insisting that animal behaviour is instead the result of God's "supereminent reason".[24]

The question of the origin of instinct is beyond the scope of the present article, but views concerning its nature obviously played a role in arguments concerning animal soul. In 1706 the Arminian theologian and journalist Jean Le Clerc (who denied that animals were machines), defending Cudworth's plastick nature against Pierre Bayle's accusation that it had dangerous implications,[25] affirmed that it was comparable to the scholastic notion of instinct, defined as a "blind principle" which makes animals act necessarily and in a certain order of which it is itself unaware; it enabled one to explain animal actions which otherwise could only be explained by "*connaissance*" and a soul with superior faculties to those possessed by human soul.[26] In reply, Bayle accused Le Clerc of reviving scholastic occult qualities,[27] which prompted Le Clerc to retort that there was nothing wrong with admitting the existence of unknown interior

---

22) Thomas Willis, *Two Discourses concerning the Soul of Brutes*, transl. Pordage (London, 1683), 34-38.

23) John Ray, *The Wisdom of God manifested in the works of the Creation*, 4th edition (London, 1704), 145 ff.

24) P.L.T., *A preservative against atheism and infidelity; proving the fundamental principles of natural religion, and the necessity and certainty of reveal'd religion* (London, 1706), 118.

25) On this dispute, see Stefano Brogi, "Nature plastiche e disegni divini: la polemica tra Bayle e Le Clerc", *Studi Settecenteschi*, 20 (2000), 51-88.

26) *Bibliothèque choisie*, 9 (1706), 371-378.

27) Pierre Bayle, *Réponse aux questions d'un provincial*, ch. CLXXX, in *Œuvres Diverses*, 4 vols (The Hague, 1727, repr. Hildesheim, 1966), 2: 996.

qualities of bodies.[28] Thus the notion of instinct, far from providing an uncontentious explanation of animal behaviour, was itself open to debate. Many authors preferred a different explanation for animal behaviour, and the use of the term was generally conditioned by a particular author's views on animal soul and whether animals could think.

## Animal Soul and Vital Matter

It is clear why discussions of thought and of the existence and nature of human soul inevitably involved debate as to animal soul. If one wished to preserve the immateriality of thinking substance distinct from matter, and to ground on it the demonstration of the soul's immortality, the question of animals inevitably arose. As they showed by their actions a degree not only of sensitivity but also a type of what looked like thought, the question as we have seen was how to explain their behaviour. If this thought was considered to be produced by a soul, the question of its nature arose. An immaterial soul entailed the danger of making it immortal, while a material animal soul could be extended to the human soul as well. As Ralph Cudworth wrote:

> [...] they who will attribute life, sense, cogitation, consciousness and self-enjoyment, not without some footsteps of reason many times, to blood and brains, or mere organized bodies in brutes, will never be able clearly to defend the incorporeity and immortality of humane souls, as most probably they do not intend any such thing.[29]

Cudworth, like Henry More, was particularly exercised by attempts to attribute activity to matter, which would break down the distinction between material and immaterial substance and thus undermine Christian teaching. These philosophers therefore criticized not only those who like Hobbes denied the existence of incorporeal

---

28) *Bibliothèque choisie*, 10 (1706), 409-410. See Jacques Roger, *Les Sciences de la vie dans la pensée française du XVIIIe siècle* (Paris, 1971), 424-425.
29) Ralph Cudworth, *The True Intellectual System of the Universe*, 44.

substance but also scientists who seemed to accord sensitivity or even a type of thought to matter. This was the case of Francis Glisson, Regius professor of medicine at Cambridge, who developed his conception of matter in a series of works in the 1650s and 1660s, culminating in *Tractatus... de vita naturae* (1672), which expounds a theory of energetic matter possessing life of itself. In this work Glisson denied that matter was dead and lifeless, claiming that it possessed an energetic nature and not only the capacity to move of itself but also, most importantly, "natural perception". The different vegetative, animal, or sensitive functions were accounted for by differences in organization, but all bodies have natural life. Glisson viewed substance as energy, and his conception of *biusia* undermined substance dualism. Henry More saw great danger in ascribing perception to matter, which undoubtedly seemed more threatening in the light of Spinoza's philosophy. He criticized Glisson in 1679 in a scholium to his *Ad V.C. Epistola altera*, a work that included a violent attack on Spinoza where he claimed that according life to matter ran the risk of encouraging atheism.[30] In his exchange of pamphlets with the Calvinist divine Richard Baxter, More accused the latter of adopting Glisson's philosophy and claimed "that hypothesis, if it were true, were as safe, if not a safer refuge for atheists than the mere mechanick philosophy is."[31] According to him Baxter's "admired and beloved Dr. Glisson" used the argument of self-motion in matter to prove "that there is universally life in matter".[32] In 1679 More labelled Glisson's position "biusianism" (using Glisson's term), or hylozoism,[33] which was the word used by Ralph Cudworth in his *True Intellectual System of the Universe* published in 1678. The two philosophers seem both to be reacting to what they saw as the danger implied by Glisson's conception of living matter. As we shall see, it was this crucial question of living

---

30) Henry More, *Opera omnia*, 3 vols (London, 1674-9), 2: 604-611; see John Henry, "Medicine and Pneumatology: Henry More, Richard Baxter and Francis Glisson's *Treatise on the energetic nature of substance*", *Medical History* 31 (1987), 30-32.

31) Henry More, *Two Choice and Useful Treatises*, ed. G. A. J. Rogers (London, 1997), II. 192.

32) *Two Choice and Useful Treatises*, II. 227.

33) More, *Opera omnia*, 2: 608.

matter which exercised minds at least as much as the question of machines.

The rejection of living matter was linked to the question of animal soul. For one solution to the problem of how to avoid according animals an immortal soul without turning them into machines was to stick to the theory of a material animal soul distinct from the immaterial and immortal human soul. Gassendi's distinction between an animal or sensitive soul (*anima*) and an intellectual soul (*animus*)[34] was adopted by several authors who claimed that it was a means of preserving the specificity of humans and a better way of defending religious truth than was the Cartesian hypothesis. The most outspoken defender of Gassendian Epicureanism in England, Walter Charleton, wrote for example: "the faculties of an animal (we exclude the rational faculty of man, from the sphere of our assertion) are identical with the spirits of it, i.e. the most subtile, most free, and most moveable or active part of its materials."[35] He accepted the activity of matter, seen in the numerous animal functions and reflex actions carried out without the intervention of the brain, and his defence of the immortality of the soul was based on the distinction of the immortal soul of man from the sensitive animal soul which explained all of animals' apparently rational actions. In his dialogue *The Immortality of the Human Soul demonstrated by the light of nature* (1657), one of the protagonists replies to the arguments of the materialist (called Lucretius), who points to the rationality demonstrated by animals:

> That many brute animals, especially such as are made tame and domestick, and frequently conversant with men, are conscious of their faults; daily experience doth testifie: but, that they are therefore animated with a soul capable of knowing itself, and its actions, by reflecting upon itself: seems to me to be altogether inconsequent; because, as Athanasius hath explained the reason and manner of

---

34) See Margaret J. Osler, *Divine Will and the Mechanical Philosophy: Gassendi and Descartes on Contingency and Necessity in the Created World* (Cambridge, 1994), 63ff.

35) Walter Charleton, *Physiologia Epicuro-Gassendo-Charletoniana or a Fabrick of Science Natural, upon the Hypothesis of Atoms, Founded by Epicurus, Repaired by Petrus Gassendus, Augmented by Walter Charleton* (London, 1654), ed Robert Hugh Kargon (New York and London, 1966), 272.

> that particular action in them, it doth import no more than what belongs to a meer sensitive soul.[36]

But such arguments did not convince everyone, for the claim that the sensitive soul was material opened the way for the claim, by analogy, that the same could apply to the intellectual human soul once one had admitted a type of matter capable of sensation. The consequences of belief in a sensitive animal soul are clearly spelled out by Cudworth:

> For besides the monstrosity of this latter opinion, in making two distinct souls and perceptive substances in every man, which is a thing sufficiently confuted by internal sense, it leaves us also in an absolute impossibility of proving the immortality of the rational soul, the incorporeity of any substance, and by consequence the existence of any deity distinct from the corporeal world.[37]

Arguments in favour of a separate material animal soul had been given added impetus by the works of Thomas Willis, who posited an animal soul "by which the brutes as well as men live, feel, move"[38] and who explained animal workings by means of a dual material soul accounting for all vital functions except reason in humans. He expanded on the nature and function of the animal spirits in *De anima brutorum* (1672), where the influence of Gassendi is evident. The material soul, carefully distinguished from the immaterial rational soul, is composed of the vital soul—compared to a flame in the blood, which "is animate or hath life", and "this animation is in its accension or inkindling"[39]—and the animal spirits distributed by the nerves and compared to light. Willis explains how the corporeal soul is transmitted and creates a new being:

---

[36] Walter Charleton, *The Immortality of the Human Soul Demonstrated by the Light of Nature*, ed J.M. Armistead (New York, 1985), 107-108. Athanasius, who represents Charleton, had referred to the explanations of animal behaviour provided by Descartes and Digby (p.106).

[37] Cudworth, *The True Intellectual System*, 46.

[38] Willis, *Two Discourses Concerning the Soul of Brutes*, Preface, [1].

[39] Willis, "Of the Accension of the Blood", in *Five Treatises* (London, 1681), 24-25.

> Because a certain heap of animal spirits, or most subtil atoms, or a little soul not yet inkindled, lies hid in the seminal humour ; which having gotten a fit cherishing or fire-place, and at length being inkindled from the soul of the parent acting, or endeavouring, or leaning to it, as a flame from a flame, begins to shine forth, and to unfold it self, a little before the foundations, or first ground-work of the body is lay'd.[40]

For Willis, who rejected the strict dualism of Cudworth and More, the atoms composing all material objects are active, self-moving and capable of sensation; the only difference between an insensitive and a sensitive body is that between "a thing unkindled, and a thing kindled".[41] The material animal soul, responsible for all the living being's functions, can also perceive, learn, reason in a very limited way and form certain judgements.[42] Willis's books were also particularly important for their attempts to localize brain functions and to situate mental activities in particular parts of the cerebral cortex,[43] and could therefore be used by those who wished to explain intellect in terms simply of the workings of the brain and to deny a separate immaterial soul. In fact, according to Lester King, Willis "had to whittle away at the functions of the rational soul, whose properties he relegated more and more to the animal soul",[44] indicating the difficulties of maintaining the distinct rational soul. As we shall see, comparisons of humans and animals were frequently used to support materialistic arguments denying an immaterial human soul, and for certain defenders of orthodoxy the danger seemed to lie above all in according life or sensation to matter. For such writers this danger lurked everywhere. Not everyone perceived the difference between Willis's position and that of Descartes, for the theory of

---

40) Willis, *Two Discourses Concerning the Soul of Brutes,* 29. Lester S. King considers this translation inadequate and provides his own: *The Philosophy of Medicine. The Early Eighteenth Century* (Cambridge, Mass., 1978), 139.

41) Willis, *Two Discourses Concerning the Soul of Brutes,* 33.

42) See John P. Wright, "Locke, Willis, and the Seventeenth-century Epicurean soul", in *Atoms, Pneuma and Tranquillity*, ed. Margaret J. Osler (Cambridge, 1991) 239-258; Duchesneau, *Les modèles du vivant*, 85-117.

43) See Alfred Meyer and Raymond Hierons, "On Thomas Willis's Concepts of Neurophysiology", *Medical History* 9 (1968), 1-15, 142-155.

44) King, *The Philosophy of Medicine*, 142.

material animal spirits was linked to Willis's corporeal soul. Dr Nehemiah Grew, who postulated a separate vital principle and denied that any type of organisation of inert matter could make it vital, wrote in 1701:

> Some indeed, as the acute Descartes, and after him, Dr. Willis, and others, have supposed, That a very subtile, aerial, etherial, or igneous fluid, conteined in the blood, brain, or nerves, or in them all; is the life of an animal. But if we go to the bottom of this conceit, how vain is it? [...] body, by being subtilized, can loose nothing of its corporeity.[45]

John Ray went further, lumping together Descartes, Gassendi and Willis as claiming "that the soul of brutes is material, and the whole animal, soul and body, but a meer machine." He of course took a very different position, arguing from animals' actions that they have a lower degree of reason to that of humans, and from the similarity of the physical structure of animals to that of humans that "it is very probable that they have the same sensations and perceptions with us",[46] while denying claims that in that case they must have an immortal soul.

As Pierre Bayle pointed out in his *Dictionary* article "Rorarius", if animals were accorded a material soul, there was no reason not to extend it to humans.[47] For many the only solution seemed to be to defend the necessary distinction between humans and animals, ensured by the animal-machine hypothesis which made animals unthinking automata; thus the difference between them and humans was one of kind rather than simply of degree. The arguments were rehearsed at length by Bayle in many pages of his *Dictionnaire*, which present clearly the difficulties of the question and its bearing on the problem of human soul and theology. As he writes in the article "Péréira": "*de tous les objets physiques il n'y en a point de plus abstrus, ni de plus embarrassant, que l'âme des bêtes. Les opinions*

---

45) Nehemiah Grew, *Cosmologia Sacra: or a Discourse of the Universe as it is the Creature and Kingdom of God* (London, 1701), 32.

46) John Ray, *The Wisdom of God manifested in the works of the Creation*, 4th edition (London, 1704), 61, 63.

47) Pierre Bayle, *Dictionnaire*, 16 vols (Paris 1820-24, reprint Geneva, 1969), 12: 603.

*extrêmes sur ce sujet sont ou absurdes, ou très-dangereuses; le milieu qu'on y veut garder est insoutenable.*"[48] In this article he reviews different arguments on the question, emphasizing that the Ancients believed the human soul to be material and only differing from animals' in quantity "and according to different degrees of subtlety" and that beasts can feel.[49] In his more famous article "Rorarius" he begins by referring to the difficulties involved in the Cartesian hypothesis of animal-machines despite its theological advantages:

> C'est dommage que le sentiment de M. Descartes soit si difficile à soutenir, et si éloigné de la vraisemblance; car il est d'ailleurs très avantageux à la vraie foi, et c'est l'unique raison qui empêche quelques personnes de s'en départir. Il n'est point sujet aux conséquences très dangereuses de l'opinion ordinaire.[50]

The crucial point is to maintain a distinction between human and animal souls, which scholastic philosophy cannot do, because once one allows that animals have a sensitive soul, one must accord that soul the knowledge or awareness ("*connaissance*") that it feels.[51] His discussion concentrates in fact on the difficulties involved in Aristotelian philosophy. While recognizing that it is hard to explain how pure machines can do the things that animals are capable of, he reiterates the advantages of Cartesianism for religion and the danger for the human soul of according beasts a soul that is capable of "*connaissance*". By demonstrating that not only thought, but also sensation, imagination and instinct require an incorporeal substance, Descartes provided the means of convincing everyone of the immortality of their soul and thus reinforcing proof of divine order.[52]

---

48) "Of all physical objects, there is none more abstruse or embarrassing than animal soul. Extreme opinions on this subject are either absurd or very dangerous; the middle path to which some wish to hold is unsustainable."

49) Ibid., 11: 546, 555.

50) Ibid., 12: 588-589. "It is a pity that M. Descartes's opinion is so difficult to uphold and so unlikely; for it is very advantageous to true faith, and that is the only reason preventing some people from abandoning it. It is not subject to the dangerous consequences of the common opinion."

51) Ibid., 12: 599.

52) Ibid., 12: 591. It is interesting that here Bayle includes instinct, which for other Cartesians, and even for Descartes himself, as we have seen, seems to cover mechanical actions not requiring an incorporeal substance.

This advantage is lost if we accord animals a sensitive mortal soul responsible for sensation and desire. Bayle develops the inconsistencies and contradictions in scholastic views of the soul and mocks their pretension in detailing the precise limits of animals' capacities, insisting that they cannot demonstrate that this soul is different from that of humans. This is where the danger resides, according to Bayle, who contemplates with "horror" the consequences of this position, namely that if animal souls are material and mortal, then so are human souls; this would overturn the doctrine of a future life and undermine the foundations of religion. On the other hand, if human soul is an immortal spiritual substance, then so is animal soul, a hypothesis that leads to innumerable difficulties.[53]

Amongst the writers mentioned by Bayle is Thomas Willis, whom he quotes at some length while countering his claim that there are two distinct souls in humans. He insists on the unity of human consciousness and accuses Willis of unnecessarily multiplying entities. Bayle emphasizes the need to preserve the distinction between humans and animals, and his complex analyses of atomistic philosophy and theories of active matter show his refusal of any compromise between the dualist position and complete materialism.[54] He states on several occasions that if the thinking substance is not separate from the body, then all bodies are thinking substances, or in other words sensation and thought are innate properties of matter.[55] In his article on Daniel Sennert he again points out the dangers of making no difference between human and animal soul, accusing Sennert of defending

> [...] un dogme selon lequel il est sûr que l'âme des bêtes est de même espèce que celle de l'homme. La différence de leur sort, quant à la durée, ne coule pas de la différence de leurs perfections, mais du bon plaisir du souverain maître, qui est une cause tout-à-fait externe. [...] Cette opinion est dangereuse; elle nous réduit à ne savoir que par la révélation l'immortalité de nos âmes.[56]

---

53) *Dictionnaire*, 12: 599, 601.

54) See Gianni Paganini, *Analisi della Fede e Critica della Ragione nella Filosofia di Pierre Bayle* (Florence, 1980), 390-404.

55) For example, "Dicéarque", rem. C, in Bayle, *Dictionnaire*, 5 : 507.

56) Ibid., 13: 241-242: "a dogma according to which it is certain that the animal soul is of the same species as that of man. The difference in their fate, as to duration, does

It is precisely the distinction between humans and animals that was denied by those who questioned the existence of an immaterial soul and tried to demonstrate the materiality of humans.

## Can Matter and Motion Think?

Christians who argued in this way at the turn of the eighteenth century seem to have been provoked by the extreme claims of theologians who denied that any type of matter could ever acquire the capacity to think, to the extent of denying that even God could accord it this property. This was the position of the Cambridge Platonists and was repeated by the first Boyle Lecturer, Richard Bentley, in particular in his second sermon, entitled "Matter and motion cannot think", preached in 1691. Bentley discussed the question of the animal soul and referred to the Cartesian animal-machine hypothesis but refused to enter into the controversy concerning animals. He claimed that it made little difference whether we accord animals souls or claim they are machines, as neither opinion endangered religion. The main point for Bentley was to

> [...] deny then that they have any reason or sense, if they be nothing but matter. Omnipotence itself cannot create cogitative body. And 'tis not any imperfection in the power of God, but an incapacity in the subject. The ideas of matter and thought are absolutely incompatible.

If one convinced the Cartesians that beasts have thought, sensation etc., they would abandon their opinion as "none but besotted atheists, do join the two notions together, and believe brutes to be rational or sensitive machines. They are either the one or the other."[57] These peremptory affirmations, in particular the denial that God was capable of creating thinking matter, shocked a certain number of Christians and led to various reactions, including from Locke,

---

not derive from the difference in their perfections but from the will of the sovereign master, who is a totally external cause. [...] This opinion is dangerous; it reduces us to knowing our soul's immortality only by revelation."

57) *Eight Sermons Preached at the Honorable Robert Boyle's Lecture...*, 5th ed. (Cambridge, 1724),75.

who developed in the following years his thinking and reading on the soul. One writer who reacted was "M.S.", apparently the theologian Matthew Smith, the author of a treatise read and annotated by Locke, called *A Philosophical Discourse of the Nature of Rational and Irrational Souls* (1695, republished in 1704). He rejected Cudworth's arguments concerning the immaterial soul and, in order to counter the Cartesian hypothesis of animal machines and demonstrate the immateriality and immortality of the thinking substance, posited the existence of a material spirit distinct from the immaterial soul, which is the animal soul and the lower human soul, responsible for feeling only. He therefore rejected Bentley's claim, writing:

> ... to say, none but besotted atheists believe brutes to be sensitive machins, is so notoriously false, that 'twas extreamly surprizing to see it publish'd by a man of learning and credit; but tho' he took this liberty with men, yet methinks he should have spoken more warily of almighty power than positively to assert, it could not create cogitative body, there ought to be undeniable reasons to support such an assertion, yet the substance of all his reasons amounts to no more than, this, we cannot conceive how meer matter and motion should produce sensation. [58]

His arguments are supported with quotations from the works of Thomas Willis. For Smith, in short:

> We readily acknowledge, we cannot conceive how mere matter and motion should produce sensation, but we say 'tis no good consequence, that therefore they are incapable by the power of God to do so : shall we say, 'tis impossible for matter and spirit to be united, and have mutual influences upon each other, because we cannot conceive how it should be so?

He attempts to demonstrate that while "brutes are sensible"," they are "devoid of will and reason", from which he shows "that the souls of all meer animals, and the sensitive souls of man, are corporeal" and "that the rational soul is a spiritual immaterial substance."[59] He undertakes to explain all the apparently intelligent behaviour of

58) M.S., *A Philosophical Discourse of the Nature of Rational and Irrational Souls* (London, 1695), 9-10.
59) Ibid., 11.

animals by their nature and divine providence (he does not use the word instinct) and then refers to anatomical writings to show that animal spirits, "made by the brain out of the vital spirits and the arterial blood their vehicle," are responsible for sensation. He quotes Willis to the effect that death is "an absolute extinguishment of the animal life, which like a flame expires." Those who claim, on the contrary, "that meer animals are a compound of matter and spirit, and that death is only a separation of body and spirit" ridicule the natural arguments for man's immortality as they apply equally in that case to animals, "which is against divinity and common sence." Thus those who equate animal life, or sense, with rational soul favour atheism as they encourage a belief that rationality does not prove immateriality.[60] Religion therefore, for this author, requires a distinction to be made between animals and humans on the lines of that made by Willis between the animal and the rational soul. This distinction guarantees human immortality while excluding that of animals. As he writes in conclusion, his aim is to show "the essential difference between the nature of man and other animals, the confounding whereof strikes at the whole of religion, and renders that unnecessary and man contemptible."[61]

As indicated above, this work was read and quoted by Locke, who had also attended Willis's lectures in Oxford in 1663 or 1664.[62] John Wright, who has compared Locke's view of the soul to that of Willis, bringing out both the similarities and the differences, concludes that Locke did not adopt the same two-soul doctrine solution, but seemed to favour some kind of corporeal souls or "systems of matter fitly disposed" like that described by Willis, as being responsible also for the higher human functions.[63] For Locke the difference between human and animal souls seems to be one of degree and not kind, as does the difference between animal and

60) Ibid., 17-19.

61) Ibid., 34.

62) See Wright "Locke, Willis and the Seventeenth-century Epicurean Soul", 243.

63) Ibid., 257-8. For Philippe Hamou, this is the only plausible alternative for Locke to Cartesian doctrine: "L'opinion de Locke sur la 'matière pensante'", *Methodos*, 4 (2004), translated in *John Locke*, ed. Peter Anstey, 4 vols (London, 2006), 3 : 177.

human faculties.[64] He discusses animal perception and memory (the "faculty of laying up and retaining the ideas, that are brought into the mind"), but doubts that animals can compare ideas or form complex ideas. Hence

> [...] if they have any ideas at all, and are not bare machins (as some would have them) we cannot deny them to have some reason. It seems as evident to me, that they do some of them in certain instances reason, as that they have sence ; but it is only in particular ideas, just as they receiv'd them from their senses. They are the best of them tied up within these narrow bounds, and have not (as I think) the faculty to enlarge them by any kind of abstraction.[65]

But Locke avoids pronouncing clearly on the question of the soul's materiality or immateriality, for reasons probably connected with the delicate nature of the question.

Another person who read Matthew Smith's pamphlet was one of the most vigorous and consistent opponents of materialistic conceptions of humans, the Rev. John Turner, vicar of Greenwich, who composed a series of works defending various aspects of Christian doctrine. In his 1698 *Phisico-Theological Discourse*, the chapter defending the immortality of the soul also drew on Smith's book to insist on the difference between human and animal soul, and between sensitivity and reason. He rejected the Cartesian hypothesis as "very injurious to our rightly conceiving the reasonable soul of man" and adopted the hypothesis of a material sensitive animal soul. Death is thus not a separation of matter and spirit but an "extinguishment of the animal life or vital flame", for to suppose otherwise would imply that animals have an immaterial soul like humans. According to Turner, the reason why a certain number of physicians have been atheists is

> [...] because they are able to demonstrate that sense is made by matter and motion, and therefore have carelessly concluded reason to spring from the same principle, and all our actions to be accounted for by mechanism: and those men help much

---

64) Philippe Hamou, "L'opinion de Locke sur la 'matière pensante'", 203.

65) Locke, *Essay Concerning Human Understanding*, II, ch.11, §11; ed. Peter H. Nidditch (Oxford, 1975), 160.

to the confirmation of this opinion, who assign the office of sensation to the rational soul, and allow reason to other animals; there is no adversary to religion, but will readily grant the animal life and rational soul to be the same thing; and that all animals are rational; but then he subjoins that the animal life is corporeal: and therefore concludes that rationality is no argument either of immateriality or immortality.[66]

## Henry Layton

One such writer, who however did not consider himself an "adversary to religion," was also reacting vigorously to Bentley's Boyle Lecture. Henry Layton, unlike Smith, used Willis to argue against a separate immaterial soul and abolished the distinction between humans and animals. This pious old gentleman from the North of England sparked off a violent debate with a series of pamphlets published at his own expense which do not seem to have had a very wide circulation although they did provoke refutations, including from one of the Boyle lecturers, who considered that his hypothesis "necessarily makes men mere machines".[67] John Hancock was here reacting to what seems to have been the first work published by Layton, his "quarto pamphlet of two and a half sheets" called *Observations upon a Sermon intitled, A Confutation of Atheism from the Faculties of the Soul,* Alias, *Matter and Motion cannot think: preached April 4.1692. By way of Refutation.* It was directed against Bentley's Boyle lecture, claiming that its author fails to prove the existence of an immaterial substance or soul distinct from the body. Layton defends the view "that the humane soul is a material spirit generated, growing and falling with the body, and rising again with it at the sound of the voice of the archangel, and the trump of God." He points to the material spirits that are said to exist in plants and animals, and asks why one should not also attribute human life and activity to a similar material spirit:

---

66) John Turner, *A Phisico-Theological Discourse upon the Divine Being or first cause of all Things*, (London, 1698), 84-85.

67) John Hancock in 1706: *A Defence of Natural and Revealed Religion: Being a Collection of the Sermons Preached at the Lecture Founded by the Honourable Robert Boyle Esq.*, 3 vols (London, 1739), 2: 234-239.

> [...] upon averment that all these faculties in the brutes were acted by a material spirit, and a demand why the same might not be effected amongst men: the maintainers of immateriality seem much put to it for an answer to this objection, and they are divided upon it.[68]

His own position is very clear:

> I do believe, that there is not any *particular thing* in man's composition that *thinks, argues etc*, but that it is the *man* himself, *viz.* the whole composition of soul and body by a divine and admirable contexture united, which thinks, argues, and doth all other things, which God hath given him a power and propensity to do.[69]

He criticizes Bentley for claiming that the question of whether animals are machines or have an immaterial soul is unimportant; for Layton, on the contrary, "if either of these opinions be true, the whole fabrick of nature and science would thereby be drawn into very great differences and alterations." Instead of stating that if men are machines they can have neither reason nor sense Bentley should, according to him, have said that as men have reason and sense, they cannot be bare machines. But Bentley preferred his "paradoxical and fallacious axiom," that "omnipotency it self cannot create cogitative body."[70] For Layton, animals neither are machines nor have an immaterial soul. He explains that thought is conducted by means of material spirits, which are "particles of the purest blood inflamed, glowing and lucid, irradiating the brain and all the ventricles or concavities of it, with the appendances of apprehension and memory thereunto belonging" ; they are the "active principle of life, motion, sense and understanding in man and beast."[71] This definition seems to be inspired by Willis, as is his comparison of humans with animals. He argues against Descartes or Kenelm Digby who denied all sensation to animals, saying that their position was so absurd as to make Dr. Willis think such arguers deserved no

68) Henry Layton, *Observations upon a Sermon intitled, A Confutation of Atheism from the Faculties of the Soul, Alias, Matter and Motion cannot think: preached April 4.1692. By way of Refutation* (n.p.n.d.), 5

69) Ibid., 2; emphasis Layton's.

70) Ibid., 13.

71) Ibid., 9-10.

answer.[72] For Layton the effects of disturbances of the brain on intellectual capacities demonstrate "that the brain of man is cogitative matter, or very like it" and

> [...] that the particles of blood inflamed, ascending from the heart unto the head are there further rarified and made lucid, irradiating the brain and the ventricles thereof; and that to these spirits and organs God hath imparted a discerning faculty or power, and that when these organs are sound and perfect, and the flaming particles clear, bright and lucid, they work to a perfection of degrees in the minds, reasons, and thoughts of men.[73]

In *A Search After Souls and Spiritual Operations in Man*, a reaction to the second edition of Richard Baxter's *Dying Thoughts* (1688) directed against the soul-sleepers and Stoics, which demonstrated that the soul is a substance,[74] Layton goes into more detail. He rejects Baxter's argument that the soul can subsist and act separately from the body although he concedes that we do not know exactly how God makes the human body or automaton live and move.[75] This ignorance comes mainly from the fact that we can only dissect dead bodies in which the vital flame has been extinguished, and as we observe only dead matter we imagine the need for an immaterial principle to account for its operations. Layton challenges believers in an immaterial immortal soul to solve the problems created by this doctrine, including how to explain animals' faculties and reproduction. He develops at some length arguments based on the similarities between humans and animals, listing as others had done before him the many different faculties exercised by animals. He claims that the only basis for postulating the existence of an immaterial soul in humans is our ignorance of how to explain our faculties, and continues: "they may as well devise souls for the beasts upon the same grounds: *viz*, for that they cannot give a rational or satisfactory account how, and by what particular means and powers many of their actions are framed, guided, and performed."[76]

---

72) Ibid., 6.
73) Ibid., 15.
74) *The Practical Works of Richard Baxter*, 4 vols (London, 1707), 3: 876.
75) Layton, *A Search After Souls and Spiritual Operations in Man* (n.p.n.d), 8-9.
76) Ibid., 14-15.

For Layton, "in men and beasts, it seems, there is a material spirit kindled and enflamed and thereby attenuated to so great a degree of subtility as to become invisible."[77] He claims that Henry More's arguments are all based on the statement "that no matter, or thing consisting of matter, can have a principle of motion or activity in itself, or of moving any other matter whatsoever", which is a "gross mistake".[78] As all animal faculties come from "the motions and actings of a material spirit, and the force and power of a natural flame," he then enquires: "what hinders then, but that a material spirit in man, may as well perform the same productions in body and mind, of those of his kind and species", purer and more subtle spirits producing a different degree of intellect.[79] Those who refuse this opinion contradict each other, some claiming that animals have an "immaterial self-subsisting spirit" while others like Digby compare them to machines. Layton also denies the need for future recompenses, argued by Baxter, More and Digby, and reaffirms God's freedom to order things as he wishes rather than having to act "according to the rules of moral congruity passing amongst men".[80]

Layton refers specifically to Willis's *De anima brutorum*, bought for him in an auction by a friend (he does not seem to know the work on the brain), which he believes provides arguments to support his position. For example, in his reply to a letter apparently from Henry Dodwell in 1702, he accounts for the different capacities of animals and humans by differences in organisation, explaining "it doth appear by the dissection and anatomy of brutal and humane creatures, which, if you will take pains to read Dr. Willis's learned book *De Anima Brutorum*, you will be more fully convinc'd of."[81] Although he claimed to have originally reached his position independently as a result of his own reflection, he is clearly extending Willis's "flamey soul" to humans. He even claims that Willis, like others, only maintained an immortal soul in man "in words:

77) Ibid., 26.

78) Ibid., 33.

79) Ibid., 41-42.

80) Ibid., 173, 181.

81) Layton, *Arguments and Replies in a Dispute concerning the Nature of the Humane Soul* (London, 1703), I, 11.

for all their reasons and deductions make against it", because he was afraid to declare it openly.[82]

In his *Observations upon a Short Treatise written by Mr. Timothy Manlove* (1697), Layton provides various relatively lengthy descriptions of automata. John Yolton, who considers that he is here arguing in favour of a human machine on the lines of the animal machine, claims that he "urged the automatic model for man" and was fascinated, like Descartes, with mechanical contrivances.[83] I would instead argue that Layton's point is rather different; he describes these machines in order to point out that the observers were convinced that magic or demons must be involved, considering "that such motion or action must needs be performed by the force and power of a separate intelligence."[84] His aim is to show, by analogy, that those who claim that human actions must be performed by an immaterial soul are equally mistaken. The point is not to demonstrate that "man as made by God could equally be mechanical"[85] but to show that a particular organisation of matter can think, or that "the skill and power of God, can act the fabrick of man's body by the force and energy of a material unintelligent spirit", once again in opposition to Bentley's limitation of God's power. He spells out this position more clearly in reply to a critical letter written by a neighbour in 1702, according to whom he had claimed "mechanical operations are sufficient for performing the soul's practices and power." He denies having written anything of the sort in any of his books

> [...] which refer production of all such powers in man or beast to the skill, the will and appointment of God. And that he can produce intelligence and intellect in the animal nature, by means of the spirits of their blood acting their *cephaline* organs to such purposes as he hath appointed them, without entring into the mazes of *mechanism*, or any thing that men know, can by that art be perform'd.[86]

---

82) *Second Part of a Treatise intitled a Search after Souls* (n.p.n.d), 22-23.

83) Yolton, *Thinking Matter*, 36.

84) *Observations upon a Short Treatise*, 35.

85) Yolton, *Thinking Matter*, 37.

86) *Arguments and Replies*, 52; emphasis Layton's.

In a later work he compares animals to pipes and whistles in which the wind produces a great variety of sounds according to the way they are made, with the aim of showing that small animals function in the same way as large ones and as men:

> [...] for each of these small animals have skin, bone, flesh, blood, sinews, nerves, muscles, joints, heart, brain, and breath, as well as the horse, or the elephant, or even as man himself, and act their five senses, and vital faculties, as other animals do, and have a flame of life, or glowing in their blood, which makes it circulate, active, and moving; as is done in other more large and potent animals, in a proportion and nature suitable to their beings.[87]

And "man is of the same kind with the other animals, and is compared by David to the beasts that perish utterly."[88] In arguing that the whole human thinks and not some immaterial substance attached to the material body, Layton seems generally to favour a view of a material soul on the lines of Willis's animal soul. This was not however the only possibility of explaining human thought without recourse to an immaterial soul. An alternative suggestion was that thought was an emergent property of a particular organisation of matter in the brain. This was apparently the theory favoured by William Coward, the main defender of material human beings in this period.

**William Coward**

Dr William Coward studied in Oxford, becoming Doctor of Medicine in 1687, and seems to have practised mainly in Norwich or Ipswich. He also seems to have nurtured aspirations to a literary career and made a name for himself by his works before the scandal created by his theological books and their condemnation by the House of Commons in 1704. Like Layton, he was a member of the Church of England, although apparently more critical of it than the old Yorkshire gentleman, tending to the criticism of priestcraft typical of the "true" or "country" whigs and perhaps, despite his

---

87) *A Search after Souls*, II (1706), 173-174.

88) Ibid., 175. The comparison is perhaps inspired by Willis.

denials, leaning toward Socinianism.[89] He published a series of works on the soul (in addition to medical and literary works), beginning with *Second Thoughts concerning Human Soul* in 1702 and publishing subsequent replies to the immediate criticism his work aroused, which further developed his views: *Farther Thoughts concerning Human Soul, in Defence of Second Thoughts* (1703), *The Grand Essay, or a Vindication of Reason and Religion, against Impostures of Philosophy* (1704), *The Just Scrutiny: or a Serious Enquiry into the Modern Notions of the Soul* (probably 1706). *Second Thoughts* begins with a defence of individual judgement based on evidence and reason, followed by detailed criticism of Aristotelian philosophy and scholastic arguments for the soul's immateriality and immortality, including mockery of those who place it in the mouth of the stomach, the pineal gland, or the heart. He then expounds "the true notion of human soul", making clear that he is aware of the dangerous company he is keeping:

> If I approve of or concur in the opinion of the learned Mr *Hobbs* of *Malmesbury*, that it is a very odd notion to call any created being an *immaterial substance*, I know I shall be received upon the very account of his name both with censure and prejudice; yet, that there are *spiritual beings*, whose constituent parts (as I may so call them, in order to be the more plainly understood, wthout any philosophical cavil about the word part) are very difficult to be comprehended by our weak understandings, I make no more doubt of than that there is a *sun* and *moon* in the *firmament*. Altho' again I must plainly declare, that my sentiments differ very much from the learned Dr. *Hen. More*, who tells us that the notion of spirit is as easy to be comprehended, as that of matter.[90]

This passage shows that, like Layton, Coward opposed theology based on the necessary existence of an immaterial substance; he too was reacting to the arguments of people like Kenelm Digby, Henry More and Ralph Cudworth. He shared Hobbes's belief in the resurrection and last judgement, which did not require an immaterial soul continuing after the body's death. The doctrine of such a soul was according to him both against the teachings of scripture and

89) For more details, see Thomson, *Bodies of Thought*, 104-117.

90) William Coward, *Second Thoughts,* 82; emphasis Coward's.

incapable of demonstration. Coward based his own definition of soul on the Bible: "the soul is a breath originally infus'd by God into insensible matter, by which it lives and exerts sense and reason."[91] It is therefore simply life or "a continued circulation of the blood".[92] The mind is not a substance but "a thinking, reasoning faculty" and thought is produced in the brain by the animal spirits, which are the purest part of the blood, but not a material soul (here he cites Daniel Sennert and Walter Charleton).[93] While he does not take issue with those who believe the soul to be fire, he prefers to discribe it as "an *afflatus* or *spiritus*, or breath, or power actuating matter as briskly (if you please) as fire would do, were it in the body."[94] Like Layton, Coward frequently bases his demonstration on the similarities between humans and animals, claiming that animals, like humans, have reason, defined as "an operation, or *impetus* of the understanding or mind, by which every animal is excited to obtain by divers ways the end he proposes to himself, according to the best of his judgment."[95] He refuses to limit this capacity to humans, claiming that it "necessarily follows life and sensation". He therefore rejects the notion of natural instinct as "giving us not the least light to comprehend any philosophical truth."[96] His critic John Broughton remarked on this that while Coward rejects natural instinct as meaningless, he himself introduces an occult quality with his definition of soul as a "power inherent in matter".[97] What others attribute to natural instinct Coward calls "natural reasoning", because instead of behaving in a stereotyped fashion, animals clearly adapt their behaviour to a particular aim. He likewise rejects the explanation of their behaviour in terms of a sensitive soul, whose existence he denies. The comparison with animals is in fact an important part of Coward's argument against the existence of an immaterial soul;

---

91) Ibid., 90.

92) Ibid., 107.

93) Ibid., 103-105, 112.

94) Ibid., 249.

95) Ibid., 97-98.

96) Ibid., 148.

97) John Broughton, *Psychologia, or an account of the nature of the rational soul in two parts* (London, 1703), 398.

he writes that the operations which result from life and are necessary properties belonging to it

> [...] are the acts of reasoning, understanding, the will, and memory, and the like, usually said to distinguish man from beast, and made the specific difference, and indeed, I think, we need not recur to a substantial immaterial soul to be the foundation of such operations, seeing they appear even in brutes, tho' in a lesser degree.[98]

In the same way as humans and animals differ in their understanding, so humans differ from each other, depending on both physical constitution and education, which he claims would not be the case if it were the production of an immaterial soul. The difference between humans and animals is therefore gradual, not specific. He also uses reasoning from like to like to demonstrate what he calls "the mortality of the whole man", basing his demonstration of the similarity between humans and animals on the text of the Bible.[99] Thus if we postulate that the existence of reason in humans demonstrates the existence of an immaterial soul, we must accept the same conclusion for animals. Like others, he provides examples of animal intelligence to show that they are not essentially different from humans. He even interprets some examples given by the defenders of animal machines to show that the actions described imply in fact a recollection of ideas; so animals possess all human intellectual capacities, having "the power of forming ideas and immaterial conceptions, and also exerting reflex actions" and will.[100] He concludes

> [...] that there is in several things a manifest parity of nature between man and a brute; tho' the perfection be supereminent in man; and that the grand characteristick, viz. reflex actions will argue as much an immaterial spirit or soul in brutes, as it will in man, if allow'd to be the only foundation of distinction between a rational and brutal soul.[101]

---

98) Coward, *Second Thoughts*, 114.
99) Ibid., 162, 156.
100) Ibid., 171-172.
101) Ibid., 175.

To explain their inferior reasoning capacities he adopts Locke's argument that speech is necessary for thought, concluding that

> ... perfect ratiocination seems to be not the natural result of an immaterial soul, but only the product of conversation, necessarily attending creatures wherein is the breath of life, and improvable to the utmost degree, where such creatures are capable of speech and conversation.[102]

The comparison of humans and animals is used throughout his various works in defence of his opinions, as when he points out that if, as his opponents claim, self-motion requires an immaterial soul, then animals require one as well as humans, or when he argues from sensation: "if it denote an immaterial substance in man, why not in brutes? Whom nothing but a *lusus ingenij* ever made mere atomata?"[103] Unlike Layton he does not refer here to Willis, who does not even appear in the list of "names of such persons and books as the author had an occasion materially to recite or consult" at the end of the work. This omission can perhaps be explained by the fact that he did not adopt Willis's theory of a material animal soul, even though one of his critics, William Nichols, linked his opinion to those of Gassendi and Willis; Nichols called it a "fancy" and inquired, "what can a brisk fiery part of the blood contribute to sensation?"[104] Instead Coward attempted to provide an explanation of thought by movement in the brain; it is "the result of certain effluviums from the brain, raised and continued by a perpetual circulation or rotation of ideas thereon impressed, as by God originally so ordained."[105] In reply to the theologian John Broughton's query as to how the material animal spirits can produce thought, Coward compared the violent motion of the spirits shut up in the skull, which prevents the motion being exhausted, to that of

102) Ibid., 2 inserted between 176 and 177.

103) W.C.M.D.C.M.L.C., *The Grand Essay, or a Vindication of Reason and Religion, against Impostures of Philosophy* (London, 1704), 165.

104) Nichols, *A Conference with a Theist*, I. 257.

105) W.C.M.D.C.M.L.C., *The Grand Essay*, 128-129.

> [...] pyrotecnical preparations of spirit of wine, which altho' highly rectifyed, yet if it be contained closely in a circulatory vessel, or bolt-head hermetically sealed, shall not appear to have wasted, or exhaled the 100. part by the agitation of the fire. Thus the scull, and exterior ligaments of it, wrap up, and so envelop the brain that but very few animal spirits exhale through the porosity of it, so as to make any considerable *dispendium*, or loss, but they pass and repass, by a continual circulation of the blood.[106]

Although his description resembles Willis's in *De Fermentatione*,[107] he does not adopt the theory of a material soul spread throughout the body; instead he tries to explain how matter thinks by a particular organization in the brain using the motion of the spirits, although his conception of this appears to be rather confused. But he clearly followed Francis Glisson's conception of energetic matter, for he referred to the Glisson's *De substantiæ vi energetica* and quoted his proof that substance moves by an internal principle. He claimed that this would help to explain "that philosophical doctrine I shall hereafter insist on viz. *that all substance*, or materia informata, *has in it a principle of self-motion, and would always exert it self, did not the almighty restrain it to preserve the due frame and order of the universe*", although he did not accept Glisson's distinction between spiritual and material substance.[108] Coward's essential argument was to deny the existence of immaterial substance and, as he states on the title page of his *Grand Essay*, "that all matter has originally created in it, a principle of internal self-motion" and "that matter and motion *must* be the foundation of *thought* in men and brutes."

The works of Coward elicited a large number of responses, including denunciations from the pulpit and attempts at censorship; although a campaign to condemn him in the Lower House of the Convocation of the Church of England failed, he was called to the House of Commons to answer charges against his first two books, which were condemned to be burned by the common hangman.

---

106) Ibid., 133.

107) Thomas Willis, *The Remaining Medical Works of that Famous and renowned Physician Thomas Willis*, Englished by S.P. Esq. (London, 1681) 14-15.

108) Coward, *The Grand Essay*, 43-46; emphasis Coward's.

Several of the authors of the books and pamphlets written against him brought up the question of the link between humans and animals, with the aim above all of denying the similarities between the two. Rev. John Turner, whose views on animal soul we have already seen, criticized Coward's notion, that men "died like the beasts, without any future recompence at all".[109] William Nichols took a very different position from that defended by Turner, denying that the souls of brutes were nothing but refined matter. He preferred a sort of immaterial principle within animals or a sensitive soul:

> Neither could I ever subscribe to their opinion, which makes the souls of brutes to be nothing but a subtile active fiery part of the blood, as Gassendus, and Dr Willis. For, after a full consideration of this fancy, a man will be as soon persuaded that a soul is made of gun-powder. For what can a brisk fiery part of the blood contribute to sensation? [...] For it is the pure almighty will, and good pleasure of God, that gives the souls of men a capacity of an immortal continuance; and if he pleases, he can deal out his bounty more scantily to brutes in giving them souls, though immaterial, but of a temporary permanence.[110]

This immaterial substance need not be immortal, for he denies that everything that is immaterial is necessarily immortal. This disagreement between Coward's opponents demonstrates the thorny nature of the problem. The main question continued to be whether to accord feeling to matter, and if one could not, then how to avoid giving animals an immortal soul. Rev. John Broughton, chaplain to the Duke of Marlborough, who included criticism of Locke in his reply to Coward's *Second Thoughts*, countered the objection based on the similarity with animals "which, I know not why, has been looked upon by some, as the most considerable." He took account of the differences amongst defenders of the human soul's immortality, admitting that it was well known that they differed on animal soul: "Some thinking it purely material, and mechanically acted; others that they have sensation, and in some degree thought; and

---

109) John Turner, *A Farther vindication of the Soul's Separate Existence and Immortality. In Answer to Dr. C—'s Farther Thoughts upon his Second Thoughts concerning Human Soul* (London, 1703), 49.

110) Nichols, *A Conference with a Theist*, II, 257-258.

consequently are acted by some immaterial substance." In either case, no conclusions can be drawn concerning humans: "if brutes are machins, the case will admit of no parallel: if they are not, the parallel cannot be carry'd on to the detriment of man's immortality."[111] For he confesses our ignorance concerning God's plans for animals, and insists that this has no bearing on human immortality.

The debate was reignited by the furore aroused by the non-juring theologian Henry Dodwell's *Epistolary Discourse* (1706) demonstrating that the soul was naturally mortal and acquired immortality only by baptism. Samuel Clarke's reply to Dodwell elicited a letter from Locke's friend, the freethinker Anthony Collins, only too pleased to exploit the disagreements amongst theologians, and led to an exchange of pamphlets on the question of the soul between Clarke and Collins. Collins's fifth objection raised the problem of animal soul; he claimed that if one argues that all sensible creatures must have an immaterial soul, the only way to avoid admitting an immortal soul in animals was "1. That all those creatures (which I call sensible) are only meer machines: or, 2. That their souls shall be annihilated upon the dissolution of their bodies", both of which could apply to humans as well. In reply, Clarke denied that one must necessarily suppose the same fate for animals and humans and insisted on God's omnipotence: he "may, without any very great difficulty, be supposed to have more ways of disposing of his creatures than we are at present let into the secret of."[112] Coward, who also reacted to Clarke's letter to Dodwell, mischievously claimed that in a private conversation Clarke had in fact accepted the idea of immortal animal souls. He claimed that Clarke had told him: "that you did not know but that the souls of brutes might be rais'd at a general resurrection, as well as human souls, in order to be transmitted to some peculiar habitations, perhaps in Saturn, Mars, or Jupiter, etc., or words to this effect."[113] Collins, in a similar vein,

---

111) Broughton, *Psychologia,* 232, 234.

112) Samuel Clarke, *The Works*, 4 vols (London, 1738), 3: 752-753, 763.

113) W.C.M.D., *The Just Scrutiny, Or, a Serious Enquiry into the Modern Notions of the Soul* (London, n.d.), 97.

attempted to show that Clarke could not answer the objections rising from the similarity between humans and other animals:

> [...] since Mr Clarke does allow, that all the sensible creatures in the universe are not mere machines, by acknowledging *that they have all something in them that is immaterial*, I desire him to tell me, consistently with the principles of his foregoing argument for the natural immortality of the thinking immaterial being in us, what else remains reasonable, but that they be supposed sometime or other to be annihilated, or else be supposed capable of eternal happiness as well as man.[114]

And elsewhere: "I cannot conceive it possible for all the numerous sensible creatures which microscopical observations do discover to us to exist in all parts of the universe, even in our own bodies, and the liquors that we drink, to be proper subjects of eternal rewards and punishments."[115] In reply, Clarke accused Collins of turning humans into machines with no free will, on the lines of Descartes's animals. If thinking is a mode of motion, then there is no liberty "but absolute necessity, such as the motion of a clock or watch is determined by",[116] and there can be no religion and no future state of rewards and punishments, as Clarke insisted again in his "Fourth Defence", which brought the exchange to an end. As we can see here, the question of machines, animal or human, was closely linked to that of necessity, defended by Collins. This was essentially the meaning given by La Mettrie to the claim that man is a machine.[117]

The theological arguments concerning the nature of animal soul and the question of divine justice rumbled on intermittently after this particular polemic ended with Clarke's "Fourth Defence" but the acute period of the debate was over. The preceding analysis has brought out the main issues at stake and the rôle played by the comparison between humans and other animals. While those who wished to abolish the distinction between humans and the other

---

114) Clarke, *Works*, 3: 777; emphasis Collins's.

115) Ibid., 3: 810-811.

116) Ibid., 3: 851.

117) On this question, which is outside the scope of the present study, see A. Thomson, "Déterminisme et passions", in *Matérialisme et passions*, ed. Pierre-François Moreau and Ann Thomson (Lyon, 2004), 79-95.

animals were frequently accused of reducing humans to machines, in fact the defenders of orthodoxy were aware that the arguments were more complex. The crucial distinction between humans and animals was as necessary to ensure that humans were not viewed as machines (with the changing implications of what that implied in the early eighteenth century) as it was to avoid according animals an immortal soul, even if some theologians did not seem afraid of this latter conclusion. The solution of according beasts some type of material soul in order to safeguard the distinct immaterial human soul also entailed in the opinion of other theologians the risk, once the possibility of sensitive matter had been accepted, of admitting thinking matter and endangering the immortal human soul. The crucial question, as we have seen, was in fact the nature of matter; whether one insisted on the inertness of matter or on the contrary admitted the existence of a type of living sensitive matter, seemingly intractable theological problems were raised which could undermine fundamental Christian beliefs. These problems crystallised around the nature of animals and the frontier between humans and other animals. The denial of any clear dividing line between the two and the claim that humans are simply more perfectly organised animals became a staple of materialistic arguments, increasingly heard in the eighteenth century and expressed in its most uncompromising form in the works of La Mettrie and Denis Diderot.

# Endowed Molecules and Emergent Organization: The Maupertuis-Diderot Debate

**Charles T. Wolfe**
*University of Sydney**

**Abstract**

In his *Système de la nature ou Essai sur les corps organisés* (originally published in Latin in 1751 as *Dissertatio inauguralis metaphysica de universali naturae systemate*, under the pseudonym Dr Baumann), Pierre-Louis Moreau de Maupertuis, President of the Berlin Academy of Sciences and a natural philosopher with a strong interest in the modes of transmission of 'genetic' information, described living *minima* which he termed molecules, "endowed with desire, memory and intelligence." Now, Maupertuis was a Leibnizian of sorts; his molecules possessed higher-level, 'mental' properties, recalling La Mettrie's statement in *L'Homme-Machine*, that Leibnizians have "rather spiritualized matter than materialized the soul." But Maupertuis also debated this issue with Diderot, who critiqued this theory in the additions to his 1753 *Pensées sur l'interprétation de la nature*. Where Maupertuis attributes higher-level properties to his living *minima*, Diderot argues that these can only be 'organizational', i.e., properties of the whole. At issue here is the degree of commitment to a form of materialism.

**Keywords**

emergence, materialism, molecule, monad, organisation, Diderot, Maupertuis

In early modern discussions of the nature, function and operation of organic bodies (animal or human), including comparisons of these with various sorts of machines, both real and imagined, it is extremely rare to find distinctions between the two being made on the basis of ontological claims about 'Life' itself or the nature of the frontier separating the living from the non-living, animate matter

* Unit for History and Philosophy of Science, University of Sydney, Carslaw F07, Sydney NSW 2006, Australia (c.wolfe@usyd.edu.au).

from inanimate matter. There are accounts of Life as 'vital heat' or fermentation (as in Fernel), as deriving from animal spirits or fibres, and so forth, but almost no explicit claims about the difference between Life and its opposite (with the notable exception of Stahl in his polemic with Leibniz). However, there is a marked shift from discussions of the 'minimum' as a mathematical, physical or metaphysical point to discussions of what we might call 'living *minima*', the smallest constituents of organic life—from the metaphysics of Bruno and Leibniz to the bio-metaphysics of Gassendi, La Mettrie, Maupertuis and Diderot. (Leibniz himself did not speak of 'living minima' and indeed avoided describing atoms of substance as 'living', since Life is a property of corporeal entities and conversely, minima are not alive; yet, as I discuss below, his doctrine of the monad was frequently read in the eighteenth century as providing a basis for a vital materialism, or at least a biologically motivated metaphysics.[1]) In the seventeenth century the term 'molecule' emerges as a name for these living *minima*, and this essay examines a particular debate on this sense of 'molecule' and vital properties.

The term *molecule* is derived from the diminutive of the Latin *moles*, or mass (in Latin, a *molecula*). Pierre Gassendi used the term *molecula* synonymously with the Lucretian term *semina rerum*, with the difference that for Lucretius these 'seeds' were simply atoms, whereas for Gassendi they are composites or compounds of atoms.[2] Indeed, one should not project the concept of molecule backwards onto early atomism (as for instance an equivalent of *concilium*), since 'molecule' is the notion of a *substantial individual* composed of atoms.[3] Molecules emerge in the chemical vocabulary of the

---

1) Georges Canguilhem, "Note sur les rapports de la théorie cellulaire et de la philosophie de Leibniz, " Appendix II in *La connaissance de la vie*, revised edition (Paris, 1980), 188.

2) Olivier R. Bloch, *La Philosophie de Gassendi. Nominalisme, matérialisme et métaphysique* (The Hague, 1971), 252 n.75.

3) Henk Kubbinga, *L'histoire du concept de « molécule »* (Paris, 2001), 109. For the impact of *semina rerum* on early modern matter theory, see Hiro Hirai's *Le concept de semence dans les théories de la matière à la Renaissance de Ficin à Gassendi* (Turnhout, 2005). On the meaning of 'molecule' in seventeenth-century chemistry, matter theory and philosophy, see Bloch, *La Philosophie de Gassendi*, 252-259; Antonio Clericuzio,

1630s-1640s; while Gassendi appears to have been the first to use the term, he was preceded in articulating a molecular theory of matter by Isaac Beeckman (1588-1637), in notebooks written between 1614 and 1629 (particularly an entry of 1620 discovered by Henk Kubbinga[4]), where Beeckman's technical term is *homogenea physica*. But for our purposes it is significant that Gassendi explicitly addresses the 'property' question, that is, *do the molecules or 'moleculae' possess sensitive or vital properties?*[5], even though he wavers on the issue. According to some interpreters, Gassendi's molecules do *not* borrow the properties of sensible objects, whereas others consider that his notion of molecule specifically "helps bridge the gap between sensible qualities and atoms."[6]

The appearance of new conceptual personae[7] such as the 'molecule' thus creates new worries for natural philosophy, notably the tension between the reductionist definition of minimal entities and the correspondingly anti-reductionist, *expanded* definition of the attributes of these entities (from Francis Glisson's organic fibres possessing 'appetition' to molecules possessing sensitive and indeed cognitive properties, as discussed here).[8] This double-barreled definitional

*Elements, Principles and Corpuscles: A Study of Atomism and Chemistry in the Seventeenth Century* (Dordrecht, 2001), 63-71; Kubbinga, *Histoire du concept de « molécule »*, chs. 6-8.

4) Kubbinga, "Les premières théories 'moléculaires': Isaac Beeckman (1620) et Sébastien Basson (1621)," *Revue d'histoire des sciences*, 39 (1984), 215-233, here, 219.

5) Gassendi, *Animadversiones in Decimum Librum Diogenis Laerti qui est de vita, moribus placitisque Epicuri*, with appendix, *Philosophiae Epicuri syntagma* (Lyon, 1649), e.g. 398.

6) Compare Bloch, *La Philosophie de Gassendi*, 253 to Clericuzio, *Elements, Principles and Corpuscles*, 65.

7) I borrow this term ("personnage conceptuel") from Gilles Deleuze & Félix Guattari, *Qu'est-ce-que la philosophie?* (Paris, 1991), ch. 3. A conceptual persona can be the cogito for Descartes, Socrates for Plato, or the Antinomies for Kant, but can also be a highly overdetermined, idiosyncratic sense of 'atom', 'molecule' or 'organism'. Put differently, the molecule for Maupertuis or Diderot plays a very different role than it does in the emergence of, e.g., molecular biology as discussed in J.C. Speakman, "The Molecule—The Evolution of a Concept," *Philosophical Journal*, 2 (1965), 55-75.

8) That the term 'molecule' could carry with it unwanted reductionist baggage is evidenced by Goethe's refusal to employ it in his translation of Diderot's *Neveu de Rameau*; see Roland Eluerd, "Note sur fibre et molécule dans *Le Neveu de Rameau*,"

and indeed, ontological problem was at the heart of a debate between two prominent figures of the Enlightenment, Pierre-Louis Moreau de Maupertuis (1698–1759) and Denis Diderot (1713-1784). Maupertuis was a major scientific figure (member of the Académie Royale des Sciences from the age of 25, its director between 1742 and 1745, and most famously President from 1746 onwards of the Berlin Academy of Sciences, appointed by Frederick II) who we primarily associate today with the principle of least action, his expedition to Lapland, which provided experimental confirmation of key tenets of Newtonian physics, and to some extent his reflections on generation and species in works such as the *Dissertation physique à l'occasion du Nègre Blanc* (1744) and its expanded version, the *Vénus physique* (1745).[9] Diderot was the chief editor of the *Encyclopédie*, a translator of medical works, author of several mathematical essays and, significantly, a regular presence at Guillaume-François Rouelle's chemistry lectures at the Jardin du Roi (now the Jardin des Plantes), between 1754 and 1757.[10] I shall not discuss the overlapping biographies of Maupertuis and Diderot further here.[11]

Their debate centred precisely on the nature of 'molecules' and their minimal or expanded definition. The textual history of the debate runs as follows: in 1751, Maupertuis published (supposedly

*L'Information Grammaticale*, 52 (1992), 14-16 (thanks to Susan Bernofsky for this example).

9) Much of what we know of Maupertuis' biography comes from his friend Laurent Angliviel de La Beaumelle's admittedly rather hagiographical *Vie de Maupertuis* (Paris, 1856), written shortly after Maupertuis' death, but published posthumously. The recent studies by Beeson, Terrall and Shank (cited below) are all of high quality, with different emphases.

10) Diderot actually served as secretary during many of the lectures, taking notes and even preparing some of the lectures for Rouelle, which were first published as *Introduction à la chymie, manuscrit inédit de Diderot publié avec une notice sur les cours de Rouelle*, ed. Charles Henry (Paris, 1887); now available as *Cours de chimie de Mr Rouelle*, in Diderot, *Œuvres complètes*, ed. H. Dieckmann, J. Proust & J. Varloot (Paris, 1975-), vol. IX.

11) A recent discovery worth noting, however, is a letter from Diderot to Maupertuis accompanying a copy of his *Lettre sur les aveugles*, in 1749. It was published in Anne-Marie Chouillet, "Trois lettres inédites de Diderot," *Recherches sur Diderot et l'Encyclopédie*, 11:1 (1991), 8-18.

in Erlangen—actually in Berlin) a Latin treatise entitled *Dissertatio inauguralis metaphysica de universali naturae systemate*, under the pseudonym Dr Baumann.[12] In 1754, he translated the text into French and published it, now with a more specifically 'biological' title: *Essai sur la formation des corps organisés* (the translation was signed Abbé Trublet). The final version of the text appeared in French with a title closer to the original, *Système de la nature*, in the 1756 edition of Maupertuis' *Œuvres*. Diderot critically discussed the 'Erlangen dissertation' (and outed its author) in the second edition of his *Pensées sur l'interprétation de la nature* in 1754 (the first edition had appeared one year earlier), in sections L-LI. Maupertuis replied to Diderot's criticisms in a *Réponse aux objections de M. Diderot* included in the third and final version of his essay, in his 1756 *Œuvres*.[13]

In his *System of Nature or Essay on Organized Bodies*, Maupertuis asks his readers to imagine a molecule "endowed with desire, aversion and memory."[14] The debate between Maupertuis and Diderot concerning the nature of what they termed 'molecule' is in fact a debate over "endowment" or "attribution" of properties: should it be applied to the element or the organizational whole? On the surface, this exchange or polemic is also a case of the two authors trading accusations of atheism with each other (for the definition of matter as possessing dynamic, organizational, indeed 'intellective' properties is of course a dangerous one). But deeper down it is also an opportunity for us to witness a moment of self-delimitation

---

12) This was also the basis for an anonymous German translation, *Versuch von der Bildung der Körpers* (Leipzig, 1761), now very rare; cf. Mary Terrall, *The Man Who Flattened the Earth: Maupertuis and the Sciences in the Enlightenment* (Chicago, 2002), 322n.

13) The edition used is Pierre-Louis Moreau de Maupertuis, *Système de la nature. Essai sur la formation des corps organisés* and *Réponse aux objections de M. Diderot*, in *Œuvres*, 2 vols. (Lyon, 1756). The *Système* and the *Réponse* are in vol. 2; hereafter cited as *Système* followed by section number (in Roman numerals), and *Réponse*, followed by page number. All translations are mine unless otherwise indicated.

14) "douée de désir, d'aversion, et de mémoire" (*Système*, § XIV); the term 'endowed' appears in §§ XXXI and LXVI. In sections XIX and XXVII Maupertuis uses instead the term "accorder": matter was "granted" properties. At the end of the book he returns to the language of endowment, speaking of "originarily endowed elements" (§ LIV, 173).

of materialism. In the early twentieth century, Paul Janet, one of the first commentators to focus on Diderot philosophically, followed a hint of La Mettrie's, for whom the Leibnizians, "with their *Monads*, … spiritualized matter rather than materializing the soul" ("Ils ont plutôt spiritualisé la matière, que matérialisé l'âme"[15]) and described Maupertuis' 'endowment strategy' as a disguised form of spiritualism. Janet added that Diderot's apparent rebuttal was merely a strategic (one might say 'ideological') one, and that metaphysically his position is actually no different from Maupertuis'. Janet opposes the materialism of the atomists, in which all qualitative changes in the universe are the product of the encounter and combination of primary elements whose essential properties are extension and solidity, to the expanded materialism of Diderot, which he identifies with Maupertuis' endowed molecules, given that in both Diderot and Maupertuis matter possesses additional, mental properties which Janet traces back to Leibniz.[16]

I will briefly examine the Newtonian and especially Leibnizian aspects and background of Maupertuis' thesis, given that his concept of molecule has been described as a 'materialization of the monad', then turn to the core of his discussion of molecules and their properties, after which I consider Diderot's challenge to Maupertuis, that higher-level properties belong to the level of *organisation* (the overall physiological structure) rather than that of the *element*. But, as I discuss in closing, their debate involves significant interpretive commitments with respect to materialism, as Maupertuis' project, far from being simplistic, was motivated by the question of how the existence of a living entity could be grounded other than by an appeal to the mystical effects of a life-force; an invocation to the equilibrium existing between the workings of the macrocosm and the microcosm; or a mechanistic reduction to properties such as size, shape and motion? This is where the strategy of the endowed molecule comes into play, the attribution of a self-organizing activity

---

15) Julien Offray de La Mettrie, *L'Homme-Machine* [1748], ed. A. Vartanian (Princeton, 1960), 149.

16) Paul Janet, "La philosophie de Diderot. le dernier mot d'un matérialiste," *The Nineteenth Century*, IX (April 1881), 697.

to the molecule, in which "the materials are the workers themselves."[17] Hence the debate between Maupertuis and Diderot on the status of 'molecules' also highlights some conceptual decisions concerning materialism, emergence and reduction in the context of eighteenth-century proto-biology and its metaphysical commitments.[18]

## 1. The Newtonian and Leibnizian Context

In order to properly locate the debate between Maupertuis and Diderot, we have to first provide some elements regarding the Newtonian and Leibnizian context, as these large-scale programs in natural philosophy also had direct influence on the formulation of positions concerning generation or development, and reflections on the nature of living beings. It is well known that there was a confounding variety of Newtonianisms operative in the eighteenth century. According to the "taxonomy" suggested by Robert Schofield,[19] France alone comprised three different strands of 'Leibnizian Newtonianism', those of Maupertuis, Fontenelle and Voltaire, which share a characteristic interpretation of the monad as subject and singularity, as distinguished e.g. from the 'Baconian Newtonianism' which was predominant in the Netherlands with Boerhaave and Musschenbroek, characterized by an emphasis on force and regularity.[20] Maupertuis is both a major figure of French Newtonianism (his early *Discours sur les différentes figures des astres* represents the first public defense

---

17) Maupertuis, *Système*, § LXI, 180.

18) I discuss the relation between certain projects in the 'proto-biology' of the mid-eighteenth century, notably natural history in Diderot and Buffon, and ontological considerations (chiefly materialism) in "'Cabinet d'Histoire Naturelle', or: The Interplay of Nature and Artifice in Diderot's Naturalism," *Perspectives on Science*, 17 (2009), 58-77.

19) R.E. Schofield, "An Evolutionary Taxonomy of Eighteenth-Century Newtonianisms," *Studies in Eighteenth-Century Culture*, 7 (1978), 175-192. For a full account of Newtonianism and its offshoots in France in the eighteenth century, see the remarkable study by J.B. Shank, *The Newton Wars and the Beginning of the French Enlightenment* (Chicago, 2008).

20) Voltaire conveyed Maupertuis' Newtonian heritage, which he approved on deistic grounds, in the nickname "Sir Isaac Maupertuis" (Voltaire, *Œuvres complètes*, ed. L. Moland [Paris, 1877-1882], XXXV, 54).

of attractionism in the Paris Academy, where Cartesian ideology still dominated) and perhaps *the* main figure involved in successively grafting more and more Leibnizian elements into this framework. Diderot's more Baconian heritage is manifest in the very title of the work discussed here, *Pensées sur l'Interprétation de la Nature*, that is, *Thoughts on the Interpretation of Nature* (recall that the full title of Bacon's main work is *Novum Organum sive de interpretatione naturae*). But Newtonianism itself, whether by using gravity and attraction as an analogical basis for studying vital properties, or in the more quantitative vein of the English 'medical Newtonians' such as Archibald Pitcairne, offered various possibilities for deriving a system of generation.

However, Maupertuis explicitly states that Newtonian attraction does not sufficiently account for organic phenomena, or even "the simplest chemical operations."[21] The laws of movement are not sufficient to explain the reproduction of living beings—which is his motivation for writing the *Système*. In the earlier *Vénus physique* he had formulated the hypothesis that natural organisms were formed by attraction alone; now, in the context of an epigenetic theory, he acknowledges that the force of attraction alone cannot sufficiently account for the production of *specifically organized bodies*: "A blind, uniform attraction distributed throughout the parts of matter would not explain how these parts arrange themselves to form even the simplest organized body. . . . Why shouldn't they unite at random?"[22] That is, the same force of gravitation-attraction which controls the behavior of bodies in space governs the formation of organic bodies. However, as the force of attraction is transformed, in Maupertuis' terminology, into *affinity*, it loses the characteristics of the simple mechanical phenomenon of the Newtonian model and gains "Leibnizian" qualities (justifying Schofield's category of 'Leibnizian Newtonianism'). Maupertuis will add properties to his living minima which they would not possess in a strict Newtonian model. His

---

21) *Système*, § III, 141.

22) "Une attraction uniforme et aveugle répartie dans toutes les parties de la matière, ne saurait servir à expliquer comment ces parties s'arrangent pour former le corps dont l'organisation est la plus simple. . . . Pourquoi ne s'unissent-elles pas pêle-mêle?" (Maupertuis, *Système*, § XIV, 146-147).

interest is to see a kind of Thinking at work—an informational model of genetic transmission as the transmission of non-material 'coding' or 'content'—but our concern for now is simply to map out his arguments for the "endowed molecule."

The Newtonian context in its different forms could be hybridized with Leibnizian elements, whether as a 'vitalized' physics or a more explicitly radical materialist project, given that "both [Leibnizianism and Newtonianism] could be imagined as offering a philosophy of nature that made innate material forces in bodies the causal agent in a deterministic universe governed by impersonal, mathematical laws."[23] Leibnizian metaphysics and theories of generation thus had a great impact on eighteenth-century thought, and are frequently considered to be major influences in the formulation of Diderot's materialism among others.[24] In a pattern familiar from the controversies over Cartesian physiology (when some of Descartes' disciples such as Regius steered his system towards materialism) or thinking matter (when Locke's skeptical approach was deliberately misread as an explicitly materialist credo), Leibniz's insistence that one not confuse or blend the physical and the metaphysical levels of his system; that mechanical science and monadic metaphysics were distinct, was disregarded.[25] Thus his idea of the organism as a 'machine of nature' was turned into a biological concept: even the vitalist physician Théophile de Bordeu named monads (along with Buffon's organic molecules) in his list of the main "hypotheses on the elements of bodies,"[26] in a trend extending at least as far as Johannes Müller in the nineteenth century. To reiterate an earlier point in part, Leibniz himself does not hold that there are 'basic' organic components, since machines of nature are, after all, machines

---

23) Shank, *The Newton Wars*, 433.

24) See (with some reservations) Yvon Belaval, "Note sur Diderot et Leibniz" and "Diderot lecteur de Leibniz ?", in his *Études leibniziennes* (Paris, 1976) and for more extensive textual consideration, but unreliable interpretive claims, Claire Fauvergue, *Diderot lecteur et interprète de Leibniz* (Paris, 2006).

25) Gottfried Wilhelm Leibniz, *Philosophischen Schriften*, ed. C.I. Gerhardt, 7 vols. (reprint, Hildesheim/New York, 1978), IV, 434f., VII, 343; *Mathematische Schriften*, ed. C.I. Gerhardt, 7 vols. (reprint, Hildesheim/New York, 1962), VI, 134f., 242f.

26) Théophile de Bordeu, *Recherches sur les maladies chroniques* (Paris, 1775), 333-334.

to infinity—although this did not prevent 'vital' or otherwise materialist readings of his concept—but of course, he acknowledges that he was inspired by living beings for his concept (much as Aristotle's concept of substance is inspired by biological entities without being itself biological).[27]

On the occasion of the Berlin Academy's essay competition on monads in 1748—which was to give rise to bitter institutional infighting, as the Academy was torn between Leibnizians such as Wolff and Newtonians such as Euler, with unfortunate effects such as Maupertuis' inappropriate usage of his authority to censure the work of the Wolffian Johann Samuel König, who had linked Maupertuis' principle of least action to Leibniz[28]—there was a great deal of 'naturalized' interpretation of the monad concept, including Maupertuis but also Condillac's 1748 prize essay for the competition.[29] Maupertuis' own brief essay on the topic declares that "Monads might in principle be nothing other than the primary elements of matter, endowed with perception and force,"[30] a description that is rather close to Diderot's in his *Encyclopédie* entry "Leibnizianisme": the monad is "the real atom of nature" ("l'atome réel de la nature"[31]), itself closely related to the "sensing and living molecules" he will postulate in *Le Rêve de D'Alembert*. Hence Canguilhem could declare with some justice that "in France in the eighteenth century, it was

27) Thanks to Stephen Gaukroger and Justin E.H. Smith for helpful discussions on this point.

28) See Terrall, *The Man Who Flattened the Earth*, 292-309; Shank, *The Newton Wars*, 474-475.

29) Condillac's election to the Academy in 1749 may have owed a lot to Maupertuis, to whom he wrote on Christmas Day of that year to express his pleasure and gratitude at being elected (Condillac, *Œuvres philosophiques*, ed. G. Le Roy [Paris, 1947], II, 533). His prize essay was only properly identified and published recently: *Les monades*, mémoire présenté à l'Académie de Berlin, critical edition by Laurence Bongie, in *Studies on Voltaire and the Eighteenth Century*, 187 (1980).

30) "Les monades pourraient n'être dans leur principe que les éléments premiers de la matière, dotés de perception et de force" (Maupertuis, *Sur les monades*, Letter VIII in *Lettres de Maupertuis* [Dresden, 1752], 57). For discussion see Annie Ibrahim, "Matière inerte et matière vivante. La théorie de la perception chez Maupertuis," *Dix-huitième siècle*, 24 (1992), 95-103.

31) Diderot, *Œuvres complètes*, VII, 692.

above all through Maupertuis that the philosophy of Leibniz informed and oriented speculation concerning the growth and structure of living beings."[32]

Now, Maupertuis himself was not a self-proclaimed Leibnizian, not least for the institutional reasons alluded to above. And his own explicit contribution to the debate on monads, the eighth of his *Lettres*, entitled *Sur les monades*,[33] a 'casual' piece in the manner of Voltaire, is somewhat of an obfuscation. But one needs only to compare Maupertuis' statement quoted above, that the monad could very well be understood as the basic element of matter, endowed with perception and force, with the views of Samuel Formey, the Perpetual Secretary of the Berlin Academy, to see that Maupertuis' molecules look suspiciously like 'materializations of the monad'; as Roselyne Rey put it, "what was a principle of change within substance has become a property of living matter."[34] Formey devoted an essay to the topic of matter theory in response to some of the submissions for the Berlin Academy of Sciences prize competition on monads: the 1747 *Recherches sur les éléments de la matière*. Targetting thinkers such as the anonymous author of the *Gedancken von den Elementen der Cörper...* (Berlin, 1746), Formey argues that deliberately or not, they confuse the system of the monads with that of the divisibility of matter.[35] He insists that the genuine "elements of matter" are not accessible through microscopes[36]; the reality of these elements must be demonstrated metaphysically (like Leibniz did for Clarke). Atoms cannot be these elements because they are merely bits of extension, not simple beings ("êtres simples") possessed of a "principle of action"[37] which explains change in Nature. The experiential fact that bodies can be active or passive

---

32) Canguilhem,"Note sur les rapports," 188.

33) *Lettres de Maupertuis*, 55-57.

34) Roselyne Rey, "L'âme, le corps et le vivant," in Mirko Grmek, ed., *Histoire de la pensée médicale en Occident*, vol. 2 : *De la Renaissance aux Lumières* (Paris, 1997), 122.

35) Samuel Formey, *Recherches sur les éléments de la matière*, § II, in Formey, *Mélanges philosophiques*, vol. 1 (Leyden, 1754), 263-265.

36) Ibid., § XII, 277.

37) Ibid., § LIV, 327.

is explained, Formey continues in proper Leibnizian fashion, by their possessing such a principle, which is in fact force.

Thus Maupertuis' notion of the molecule belongs to a broader context of materialist and/or biologically oriented interpretations of Leibnizian monads. A desire to account for the phenomena of generation leads him to look for forces and properties other than basic Newtonian attraction, to account for the regular production of organized bodies, i.e., organisms.

## 2. Maupertuis' Argument

Maupertuis introduces his argument for 'endowment' in section XIV of his book:

> If we wish to be able to formulate anything about this, even if we can only conceive it through analogy, we must have recourse to a principle of intelligence, to something akin to what we call *desire, aversion, memory.* [38]

"This" is the question, why do the parts unite to form a body, and especially, why this body rather than another? Notice the admission that this can only be conceived "through analogy." Maupertuis adds later on that the analogy is intended to *complement* trials through experience, when dealing with species other than our own, and, moving towards increasingly abstract or remote cases, he explains that God did endow the smallest parts of matter with properties *like* ("semblable") what we refer to as "desire, aversion and memory" in us (it is not clear on what basis he selects these three properties, which sound faintly monadological but could also be derived from a variety of other sources).[39] Lastly, he adds that we should not worry about attributing a principle of intelligence to matter, for it is in fact not an intelligence like our own.[40]

---

[38] "Si l'on veut dire sur cela quelque chose qu'on conçoive, quoiqu'encore on ne le conçoive que sur quelque analogie, il faut avoir recours à quelque principe d'intelligence, à quelque chose de semblable à ce que nous appellons *désir*, *aversion*, *mémoire*" (*Système*, § XIV, 147, reiterated in §§ XXVIII, 155-156 and XXXI, 157-158, and again in § LXVI, 183 (with God now explicitly named as the 'donor').

[39] Ibid., § LIV, 172.

[40] Ibid., § LXII, 181.

Yet Maupertuis nonetheless fears that attributing a kind of intelligence to matter would be viewed as tantamount to materialism, precisely in the vein of Lockean thinking matter. And he was not mistaken: the attribution of generative, organizational and intellective powers to matter earned him the epithet of 'materialist' (or Spinozist, or hylozoist) in a variety of responses ranging from an anonymous review in the *Bibliothèque raisonnée* to very visible statements by the likes of Rousseau in the "Profession de foi du vicaire savoyard" (1762) and Kant in *The Only Possible Proof in Support of the Demonstration of the Existence of God* (1763).[41] After all, shortly after Maupertuis published (anonymously) the first, Latin version of the *Système de la nature*, Fontenelle famously warned in his *Théorie des tourbillons cartésiens* that attributing attraction to matter was a small step away from granting it the power to think:

> The Newtonians can say that, just as bodies only move by the will of God, it is possible that by this same will they mutually attract one another; but the difference is a major one. In the first case, God's will merely enacts an essential property of matter, its mobility, and inclines its natural indifference towards rest or motion, towards motion. But in the second case, how could bodies have a disposition to attract each other by themselves? God's will would have no relation to their nature, and would be wholly arbitrary, which is quite contrary to the order of the universe witnessed everywhere. If we grant this arbitrariness, we destroy any philosophical proof of the spirituality of the soul. *God could just as well have granted thought to matter, as attraction.*[42]

---

41) Anonymous review of the *Vénus physique*, *Bibliothèque raisonnée*, 35 (1745), 312. For Rousseau's blanket dismissal of the idea of the 'living molecule', see "Profession de foi du vicaire savoyard," in *Emile, ou de l'éducation*, *Œuvres complètes*, ed. B. Gagnebin & M. Raymond, vol. 4 (Paris, 1969), 575; Kant specifically targets Maupertuis' molecules possessing memory, desire, aversion, even if by analogy only, as a major case of the dangers of hylozoism (*Der einzig mögliche Beweisgrund zu einer Demonstration des Daseins Gottes*, as cited by John Zammito, "Kant's Early Views on Epigenesis: The role of Maupertuis," in Justin E.H. Smith, ed., *The Problem of Animal Generation in Early Modern Philosophy* [Cambridge, 2006], 317–354, here, 343).

42) Bernard Le Bovier de Fontenelle, *Théorie des tourbillons Cartésiens; avec des réflexions sur l'attraction* [1752], *Réflexions sur la théorie précédente*, § III, in *Œuvres de Fontenelle* (Paris, 1829), III, 71, emphasis mine. For brief discussion of the last sentence, see Pierre Sudaka, "L'intervention de Maupertuis dans la philosophie," in Olivier Bloch,

How does Maupertuis avoid the charge of materialism? Rather than claiming that the property of thought belongs to matter *in toto*, he makes an excursus into animal-machines and reminds the reader that the earliest (and most orthodox) theologians granted intelligence to animals. Now, he adds, thanks to the microscope, we discover smaller and smaller animals every day, which establishes that animals are aggregates (literally "heaps," "amas") of matter of varying sizes. Why not grant, then, that intelligence might be present in infinitely small parcels of matter?[43] This is sufficient to dispel theological worries, in his view, so he now turns to the properly philosophical objection of dualism, according to which body and soul are separate substances, and intelligence belongs to the soul (§ XXI, 150). Maupertuis is willing to accept the distinction between extension and thought as self-evident but he adds that these are *properties* of a *subject* "whose specific essence is unknown to us" (§ XXII, 151). If we wish to attend empirically to natural phenomena (§ XXIII, 152), specifically the formation of organized bodies, and we apply the rule of the reduction of phenomena to the smallest number of principles, we see that the Cartesian principles are insufficient (§ XXV, 154). Maupertuis quickly adds that he is not calling Genesis into doubt—Scripture adequately explains "how all these bodies emerged out of nothingness" (*ibid.*)—but rather the rules of the *preservation* of the world *after* creation (§ XXVII, 155). In order that bodies as they exist today can have formed, Maupertuis suggests that matter must possess properties other than merely physical ones: we must "grant it some degree of intelligence, desire, aversion, and memory" (§ XXVII, 155; in his response to Diderot he adds "instinct"[44]). Diderot will explicitly reject this attribution of 'mental' properties to material entities.

But notice that Maupertuis is not so much trying to engage in high metaphysics, postulating some kind of higher intelligence embedded in matter; he is endeavoring to explain *how* an organism becomes

---

ed., *Actes de la journée Maupertuis* (Paris, 1975), 61; David Beeson, *Maupertuis: An Intellectual Biography* (Oxford, 1992), 51.

43) Maupertuis, *Système*, § XVI, 148; § XVIII, 149.

44) *Réponse aux objections de M. Diderot*, 215.

what it is, how a certain quantity of matter goes through a series of transitions to become an organized body, such as a fetus.[45] In order to do so, he attributes self-organizing properties to the molecule. This act of attribution is one possible materialist approach to the question of generation or development, whereas the other emphasizes the primacy of organizational features, as we will see with Diderot; but both have in common the rejection of Cartesian matter theory, for matter in this new configuration (in which both Leibniz and Spinoza play a role, as they do explicitly for someone like John Toland[46]) is already a bearer of its own internal dispositions, starting with movement and moving upwards from there to cognitive properties. The molecule is not describable in terms of passive, Cartesian matter *qua* extension but rather active, Leibnizian matter[47]; even in Spinoza there was no recognition of a *particular potency or drive of living entities* (the conatus has nothing particularly 'vital' about it), a potential which Maupertuis, Diderot and others witnessed in Leeuwenhoek and Hartsoeker's spermatozoa, Réaumur's chrysalids, Bonnet's aphids, Trembley's polyps, or the 'animalcules' of the *anguillard*, Needham.[48]

Let us return to Maupertuis' train of argument. The notion of matter containing within it an intelligent principle actually presupposes a notion of intelligence, Maupertuis points out: a unified,

---

45) Maupertuis asks from what stage onwards the fetus possesses a soul, and wisely dismisses the question as an artificial one (§ LIX, 178).

46) John Toland, *Letters to Serena* (London, 1704), Letter V, "Motion essential to Matter," 163-239. Tristan Dagron, in his new study *Toland et Leibniz. L'invention du néo-spinozisme* (Paris, 2009), argues that Toland is less a materialist interpreter of Spinoza than he is an interlocutor of Leibniz's who develops in debate with the latter a new, 'endowed' concept of matter.

47) Jacques Roger, *Les sciences de la vie dans la pensée française au XVIII^e siècle* [Paris, third edition 1993], 479. As Cassirer put it, "In his approach to the concept of the monad Maupertuis does not attempt to conceive the fundamental units of which all natural processes consist as metaphysical points—as Leibniz had done—but as physical points. In order to arrive at these units we need not depart from the corporeal world as such, but we must enlarge the concept of matter in such a way that it does not exclude the basic facts of consciousness" (*The Philosophy of the Enlightenent*, trans. Fritz C.A. Kölln & James P. Pettegrove [Princeton, 1951], 88).

48) Cf. *Système*, § XL, 162.

indivisible notion of intelligence (he does not give a particular definition of what he means by this, but the Cartesian echoes are fairly clear, just as the monadological overtones are at other times). But why can't we see that our own intelligence actually exists in a graded set of degrees? Our levels of conscious awareness are variable, in both senses of the term: they are changing, but also able to extend over a wide spectrum ranging from sleep to wakefulness (§ LXII, 180-181). In other words, we should not worry about attributing intelligence to the elements of matter, since our own 'total' intelligence is manifold and changing. And what is the contrary view, which denies any intelligence to the constituent 'atoms' and indeed reduces the workings of the universe to atoms and chance? Epicurean atheism, the system of the "freethinkers (*libertins*) of the present day" (§ LXIV, 182-183). This is less an instance of the inextricability of biological and theological concerns in early modernity than it is a deliberate strategic move on Maupertuis' part: when advancing a controversial view about the mind, organic bodies, intelligence, liberty and necessity, and so forth, it is extremely useful to be able to counterpose this view to a recognizably 'Epicurean', 'Spinozist' or otherwise materialist position.[49] So the attribution of a form of intelligence (instinct, memory, etc.) to matter or rather the 'molecule' is not meant as a self-consciously materialist affirmation by Maupertuis (who was comfortable using Epicurean-Lucretian arguments against final causes in nature in earlier publications such as the 1750 *Essai de cosmologie*). The fact that living matter possesses a kind of potential (formative or generative, like a *Bildungstrieb* or *nisus formativus*) is chiefly for him evidence of a kind of Mind or Intelligence. Without such a principle, the body could not develop as an autonomous whole; all bodies would have been continuously developing or fully formed since Creation.

But the predominant (call it 'Cartesian') conception that intelligence has to be distinct from the properties of a material component,

[49] Anthony Collins thus distinguishes his (quite radical) determinism from the Epicurean doctrine of atoms and chance in his *Philosophical Inquiry concerning Human Liberty* (London, 1717); Joseph Priestley suggests in his *Disquisitions Relating to Matter and Spirit* (London, 1777) that compliance with substance dualism is a Catholic tenet, not part of core Christian commitments.

even a living material component such as the molecule, needs further rebuttal. According to this predominant view, our 'sense of unity', our 'inner sense' (a notion going back to Augustine[50]) cannot have a manifold, composite origin. Maupertuis elegantly suggests that when all these elements, each possessing its own desire and perception, are gathered together as a unity, the result is an *individual perception* which is stronger and more perfect than any particular elementary perception,[51] since these have lost their "particular sense of self"[52]—a kind of holism. This is why we have no direct experience of the origin of the elements of which we are composed, "our origin must be lost to us" (*ibid.*). A given quantity of elementary parts will always produce an equivalent quantity of concurrent perceptions: the same temperament, aversions, and talents ...[53]

All of this sounds rather far removed from Maupertuis the 'forerunner of genetics' or of Darwinism, as he was once known. But within the same discussion he turns to the informational aspect of this notion of 'elementary perceptions' and suggests that it explains why certain cognitive or intellectual traits of a parent are present in the mind ("soul") of a child, or why a well-bred hunting dog produces offspring who are equally good at hunting.[54] Maupertuis considers a great deal of genetic variety—he studied various cases of hereditary anomalies, from albinos to the Ruhe family in Berlin, which exhibited the condition of polydactyly (specifically six fingers on one hand) for several generations of the descendants of Elisabeth Horstmann, and he calculated the chances of it being hereditary[55]—but what it implies for him in this context is that neither chance, nor divine design are the relevant sorts of explanations for the combination of stability and variety that he witnesses therein.

---

50) Augustine, *De libero arbitrio*, II, 3, 5.

51) David Beeson sees this as a "major concession to materialism" (*Maupertuis*, 212).

52) Maupertuis, *Système*, § LIV, 172.

53) Maupertuis continues (§ LVII, 176) with a rather obscure consideration of the possibility of education (and ethics) given this determined genetic structure.

54) Ibid., § LVI, 174.

55) *Lettre* XIV, "Sur la génération des animaux," in *Œuvres*, vol. 2, p. 267; in the translation appended to Michael Hoffheimer, "Maupertuis and the Eighteenth-Century Critique of Preexistence," *Journal of the History of Biology*, 15 (1982), 142.

Granted, "the first production of any system is a miracle."[56] But God does not have the resources to produce a miracle *each time* a new body is produced. There must be a specific formal explanation for the workings of organic bodies (Maupertuis' particular concern is with biological development, not with organismic laws or the principle that distinguishes living from non-living matter). How could other systems of generation explain phenomena that he has witnessed such as resemblance through heredity, monsters, hybrids or *métis*?[57] A recurrent anomaly such as the presence of six fingers on one hand across several generations, refutes preformationism, the system according to which the child is fully contained (formed) in paternal semen, or in the maternal ovum. On Maupertuis' view, it is an accidental material arrangement which is then repeated as a pattern, until a stronger pattern replaces it.[58] But he is not specifically concerned with the *mode* in which the seminal principles are arranged. Other processes of generation may exist as well, governing the growth of "other animalcules which can be found swarming in most fluids."[59] Neither parthenogenesis nor spontaneous generation present any difficulty for his system: if there is such a thing as spontaneous generation, it does not affect Maupertuis' thesis that a new individual is produced by the arrangement of seeds (*semences*) in combination.[60] We can treat the original production as miraculous,[61] but if we consider molecules as possessing the three key properties of desire, aversion and memory, then we can understand the process by which they combine in regular, coherent ways: "In matter's initial state of fluidity, every element placed itself in

---

56) Ibid., § XXX, 157.

57) Ibid., § XXXII, 158.

58) Ibid., §§ XXXVII, 160-161, XLV, 164. Jacques Roger considers this to be "the first complete statement of generalized transformism" (*Les sciences de la vie*, 484).

59) Maupertuis, *Système*, § XL, 162.

60) Ibid., § XLVI, 165. In *most* species, these reproductive elements are to be found in certain secretions; but apparently in other species, these *éléments propres à s'unir d'une certaine manière* are to be found either *within a single individual*, or outside of the individual-to-be-produced (*ibid.*). Maupertuis refers (§ XLVII, 168) to Buffon, *Histoire naturelle* II, 8-9 for an analysis of the same fact with plants.

61) Ibid., § XXX, 157.

the most suitable location to form bodies which no longer show traces of their formation."[62]

A key feature of Maupertuis' account of genetic transmission is what he calls, self-consciously acknowledging the extent to which he cannot escape analogy in his vocabulary, the *memory* of the molecule. Is this a kind of conscious memory or more of what one might term a 'positional' memory? It is clear that for Maupertuis the 'endowed' molecules possess a kind of positional memory combined with a drive ("desire," "instinct") to combine with other molecules in structurally coherent ways. As Shirley Roe puts it, "By 'remembering' their former locations and by possessing an instinct to unite, the seminal particles would be able to arrange themselves into an embryo in the proper fashion."[63] Even the sterility of mules can be explained thus: the particles retain the memory, the 'habit' so to speak of their former location in a mare or a donkey, but their being located in the new formation (a mule) has the consequence that the organism cannot reproduce. But in fact the memory of Maupertuis' molecules is not strictly positional, as François Jacob observed: if we compare them with Buffon's "organic molecules," the memory possessed by Maupertuis' endowed molecules is indistinguishably material (spatial, positional) and psychic, whereas that possessed by Buffon's organic molecules is restrictively positional and structural.[64] Diderot himself considers that Buffon's organic molecules are easier to accept scientifically because Buffon does not attribute 'psychic' properties to them such as intelligence or memory.

In sum, Maupertuis' concept of molecules 'endowed' with higher-level properties is a distinctive claim, contrasting both with a classic atomistic model in which basic elements ("éléments bruts") lacking any form of intelligence or appetition form more complex bodies throughout the universe, in and through their chance encounters,

---

62) Ibid., § LI, 170.

63) Shirley Roe, *Matter, Life, and Generation. Eighteenth-century Embryology and the Haller-Wolff Debate* (Cambridge, 1981), 15; similar summary in James L. Larson, *Interpreting Nature: The Science of Living Form from Linnaeus to Kant* (Baltimore, 1994), 136.

64) François Jacob, *La logique du vivant. Une histoire de l'hérédité* (Paris, 1970), 92-93.

and with a voluntaristic 'design' claim in which a Supreme Being, distinct from matter, makes use of the elements of matter like an architect uses stone to make buildings. In Maupertuis' vision, the elements themselves, "*endowed with intelligence*, arrange and unite themselves in order to fulfill the aims of the Creator."[65]

## 3. Diderot's Criticisms

Diderot rejects the idea of endowed molecules, at least in its essential form as stated by Maupertuis.[66] His arguments focus on the positions expressed in sections LII, LIII, and LIV. In order to challenge Maupertuis' hypothesis, which he accepts at the level of its "empirical" benefits but not as an overall "speculative" claim, he seeks to push it as far as it can go; whether we agree with Maupertuis or Buffon, Diderot says, what is important is the extent to which these hypotheses on generation contribute to the progress of "experimental physics, rational philosophy, and the discovery and explanation of the phenomena of organization"[67]—with the latter term referring to biological structure. In addition he is ironic about Maupertuis' fears of falling into materialism (or of being taken as such).

The problem with Maupertuis, for Diderot, was that his molecules seemed to have been spiritualized (recalling La Mettrie's distinction), whereas he, Diderot, wanted to materialize the realm of the spiritual (i.e. the mental): first there is matter and motion, and gradually, through corpuscular arrangements of increasing complexity—which he terms "organization"—the phenomena or rather faculties of desire, aversion, memory, etc., are added on. Now, this seems rather mechanistic in contrast to Maupertuis; and indeed at

---

65) Maupertuis, *Système*, § LXVII, 183-184.

66) Marx W. Wartofsky, "Diderot and the Development of Materialist Monism," *Diderot Studies*, 2 (1952), 292-293 is still one of the best commentaries on the topic. (It was reprinted in Wartofsky, *Models. Representation and the Scientific Understanding*, Boston Studies in Philosophy of Science, vol. 48 [Dordrecht, 1979].)

67) *Pensées sur l'interprétation de la nature* (Paris, 1753/1754), § XII; hereafter *Pensées* followed by the section number (and page numbers in the *Œuvres complètes*).

this point Diderot introduces the "dull sensitivity" (*sensibilité sourde*) by means of which all molecules have their place, or fit into place. To be precise, Diderot allows for two properties: this rudimentary form of sensitivity, and an "automatic restlessness" (*inquiétude automate*) which leads the molecules into a variety of possible locations.[68] In combination, these two 'basic' properties are meant to account for the same variety of organic forms as Maupertuis' multiple properties (desire, aversion, intelligence etc.). Maupertuis' basic properties are more 'teleological' than Diderot's, and overall are too excessive: Diderot thinks that Maupertuis could have stopped at mass, motion—and this "mute" or "dull" sensitivity. Of course, Diderot wavers on the exact locus of this sensitivity: does it belong to matter as a whole, or only to *organized* matter?

In some texts, Diderot suggests that "sensitivity or touch is common to all beings," attributes sensitivity to the molecule, or to matter as a whole; in his entry "Leibnitzianisme" in the *Encyclopédie*, he brings together Aristotle's entelechy, Leibniz's monads, and "sensitivity as a general property of matter."[69] But in others, such as the 1765 Letter to Duclos, he denies that sensitivity can be a property of a molecule, specifically because it can only be a property of matter itself, and complicates the issue further by introducing a distinction between "inert" sensitivity and "active" sensitivity.[70] In *Le Rêve de d'Alembert*, Diderot presents the unity of the person as the result of "the gradual apposition of several sensitive molecules"[71];

---

68) Diderot, *Pensées*, § LI, in *Œuvres complètes*, IX, 84. A variety of texts of this period, whether 'chemical' (Rouelle, Venel), 'medical' (Sauvages, Bordeu) or other (Maupertuis and here Diderot) describe molecules as agitated by an internal motion, in some cases viewed as characteristic of 'Life'. Thus François Boissier de Sauvages, in his 1752 *Dissertation sur les médicaments qui affectent certaines parties du corps humain plutôt que d'autres*, explains that medicines work on the body because of the internal agitation of their molecules (in Sauvages, *Les Chefs-d'œuvre de Monsieur de Sauvages, ou Recueil de dissertations qui ont remporté le prix dans différentes Académies* [Lausanne and Lyon, 2 vols., 1770], II, 20.)

69) Diderot, *Éléments de physiologie*, in *Œuvres complètes*, XVII, 308; "Leibnitzianisme," *Encyclopédie*, IX (1765), 371.

70) "Lettre à Duclos," October 10 1765, in Diderot, *Correspondance*, ed. Georges Roth, vol. 5 (Paris, 1959), 141.

71) Diderot, *Rêve de D'Alembert*, in *Œuvres complètes*, XVII, 134.

"each sensitive molecule possessed its self before the application; but how did it lose it, and how, out of all these losses, did the consciousness of a whole result?" (*ibid.*) An answer to this question, why the molecules connect to one another, is provided by Maupertuis: there is a kind of agreement or 'contract' between them such that the perception of each element is combined with the perceptions of the others. The result, as we have seen, is "an individual perception which is stronger and more perfect than any particular elementary perception."[72]

It is in this context that Maupertuis introduces the metaphor of the bee-swarm, which will prove to be one of the most successful and evocative images of organic unity in the eighteenth century, in various renditions, some more organismic, some more mechanistic: "Thus an army seen from a distance might appear to us as a great animal; thus a bee-swarm, when the bees are assembled and united on the branch of a tree, only presents to our gaze a body lacking any resemblance to the individuals which composed it."[73] While Diderot also makes use of this image of organic unity as a composite of living parts, he emphasizes somewhat disingenuously that for Maupertuis it illustrates the 'totalizing' principle that all molecules form a whole with a unified perception—a sense of the whole (*le Tout*) which, Diderot claims, will "force [Maupertuis] to admit that the world can be God."[74] Indeed, the accusation that a strong

---

72) *Système*, § LIV, 172; compare the overtly political metaphor used in the earlier *Vénus physique*, when Maupertuis asks if instinct, "like the spirit of a Republic," is "present throughout the parts of the body, or, like in a Monarchic state, if it only belongs to one indivisible part?" (*Vénus physique*, final chapter, § II , in *Œuvres*, vol. 2, 132).

73) *Système*, § LI, 170-171. Bordeu's 1751 *Recherches anatomiques sur les glandes* also presents this image (but the book was written over several years prior to this date); the vitalist physician Ménuret de Chambaud discusses Bordeu and Ménuret's respective claims to priority in the *Encyclopédie* article "Observation"; finally, Diderot returns to the image at length in his 1769 opus *Le Rêve de d'Alembert*. For comparison between these authors' usage of the image of the bee-swarm, see Colas Duflo, "Diderot et Ménuret de Chambaud," *Recherches sur Diderot et sur l'Encyclopédie*, 34 (avril 2003), 25-44; Charles Wolfe & Motoichi Terada, "The 'Animal Economy' as Object and Program in Montpellier Vitalism," *Science in Context* 21:4 (December 2008), special issue on *Medical Vitalism in the Enlightenment*, 537-579, here, 550-554.

74) Diderot, *Pensées*, § L, in *Œuvres complètes*, IX, 82.

notion of the Whole entails the collapse of individual perceptions and thus individuality itself into a general perception was a standard argument against Spinoza, running at least from Bayle and Leibniz to German Idealism.

Diderot refers to Maupertuis' position provocatively as "the most alluring (*séduisante*) kind of materialism,"[75] referring specifically to the notion I have referred to here as 'endowed molecules'. As we have seen, he deems these molecules to be over-laden with properties, and prefers a system based on a more minimal "dull sensitivity" (which of course is simply another form of materialism, perhaps a purer one). So on the one hand Diderot radicalizes Maupertuis' position, towards an "alluring" materialism (while ironically reproaching him for falling into this trap); on the other hand he looks for a less metaphysically invested form of the same 'vital materialist' position. By seeking to force Maupertuis to take a stance on the notion of the Whole, it is possible that Diderot is trying to achieve an even more total victory for materialism: because either Maupertuis accepts it (and then Spinozism is true) or he rejects it, so that the universe is *not* an ordered whole (and then Lucretianism is true).[76]

## 4. Maupertuis' Response

From the outset[77] Diderot had declared that he intended to push Maupertuis' hypotheses to their ultimate ... or absurd conclusion. He claims that it is necessary to do so, in order to unmask the "terrible consequences" of the Erlangen Doctor's theory. Maupertuis responds that if one were not already convinced of the sincerity of Diderot's religious beliefs, one might suspect that his intention is not so much to overturn his own theory as to "himself draw these consequences he calls *terrible* from it."[78] Indeed, if we open the

---

75) *Ibid.*, § LI, in *Œuvres complètes,* IX, 84.

76) François Duchesneau appears to suggest something like this in *La physiologie des Lumières. Empirisme, modèles et théories* (The Hague, 1982), 253.

77) Diderot, *Pensées sur l'interprétation de la nature*, § L, in *Œuvres complètes*, IX, 77.

78) Maupertuis, *Réponse*, 197.

works of respected authors who are no longer living, such as Descartes or Malebranche, and apply Diderot's method of inferring consequences, the result would be that "Genesis is denied, and the freedom and power of God are nowhere to be found."[79] In any case, Maupertuis adds, all systems are fallible, because they are *finite*.[80] His retort to Diderot is that the latter's materialism is a form of systematic abstraction just like Cartesianism.[81] While this may appear merely polemical, it is nevertheless the expression of a moderate skepticism which characterizes Maupertuis' argumentation throughout his intellectual career, not just in response to Cartesianism (or materialism) but also as a way of "leveling the playing field"[82] when confronted by the competing claims of Newtonians and Leibnizians, in his twofold capacity as a natural philosopher and the President of the Berlin Academy.

But the key component in Maupertuis' response to Diderot is a notion which is equally important in Diderot's own thought, that of the Whole (*le Tout*).[83] It is because both Maupertuis and Diderot are committed to a notion of the universe as a substantial Whole that they are 'neo-Spinozists' in a broad sense (in addition to the specific, idiosyncratic sense manifest, e.g. in Diderot's article "Spinosistes" in the *Encyclopédie*, in which so-called "modern Spinozists" are described as holding a version of the theory of biological epigenesis), to use the term coined by Paul Vernière.[84] In this integrated,

---

79) *Ibid.*, 198.

80) *Ibid.*, 199-200.

81) Beeson, *Maupertuis*, 251. Maupertuis makes this point at greater length in his Letter VII, 'On Systems' (*Lettres de Maupertuis*, 48-54).

82) Shank, *The Newton Wars*, 287.

83) For an excellent brief presentation see Jean-Claude Bourdin, "Tout, le tout," in Bourdin, Annie Ibrahim *et al.*, eds., *L'Encyclopédie du Rêve de D'Alembert* (Paris, 2006), 364-368.

84) Paul Vernière introduced the influential (and controversial) category of 'neo-Spinozism' in his *Spinoza et la pensée française avant la Révolution* (Paris, 1954, 2nd edition 1982); he defines it as a form of holist materialism *founded on the life sciences* rather than on a priori metaphysical speculation (529). At least Diderot's entry "Spinosistes" corresponds to this description. Vernière describes Diderot and Maupertuis as neo-Spinozists in an editorial note to his edition of Diderot, *Œuvres philosophiques* (Paris, 1961), 229. For more extensive discussions of Diderot's usage of Spinoza and

causally closed universe, there is no particular demarcation between dead and living matter.

What is this notion of the Whole and how does it function in Maupertuis' response?

First, according to Maupertuis, Diderot is vague in his usage of terms like "the Whole."[85] If (a) the whole, *le tout*, is that which "leaves nothing beyond itself" *(ne laisse rien au delà)*, the question of whether or not the universe is a whole or not is irrelevant to the issue; if (b) a whole, *un tout*, is a regular edifice, an assemblage of proportionate parts, the previous question can be answered affirmatively or negatively.[86] If the answer is No, God's existence will not be put into question, any more than it was by pious authors such as Malebranche, for whom the universe was a heap of ruins.[87] If the answer is Yes, it does not follow that this "copulation of perceptions" necessarily extends to the entire Universe (*ibid.*). According to Diderot, all systems ultimately are founded on reasoning by analogy; this is mistaken in Maupertuis' view, since analogy will never establish truth or falsehood (a somewhat odd position to take for someone whose most significant description of the molecule explicitly relied on analogy).

Second, certain modern philosophers, following the principle that *Nature makes no leaps*, view matter as one single continuous piece (*bloc*). If Diderot meant this notion of continuity when he spoke of the Whole, one should reply that reason and experience prove that there are vacuums in Nature; bodies are merely scattered in space.[88] Maupertuis insists that regardless of the degree of continuity of the universe, any real difference between *le continu* and *le disséminé* will simply be a matter of *distance between parts*: hence

---

Spinozism, see Alexandre Métraux, "Über Denis Diderots physiologisch interpretierten Spinoza," *Studia Spinozana*, 10 (1994), 121-134 and John Zammito, "Naturalizm XVIII Wieku. Spinozyzm w Filozofiach nauki Diderota i Herdera," in *Rozum i S'wiat: Herder i filozofia XVIII, XIX i XX wieku*, eds. Marion Heinz, Maciej Potepa, Zbigniew Zwolin'ski (Warsaw, 2004), 117-146.

85) Maupertuis, *Réponse*, 204.

86) Ibid., 205.

87) Ibid., 206.

88) Ibid., 207.

his cherished elements will continue to exist. Degrees of density exist for our perception. The microscope refutes the claim of absolute continuity, since it shows distance between apparently homogeneous, compact parts.[89] (Maupertuis does not seem to have undertaken microscopic observations himself, relying instead on the reports of Needham's "infusoires.")

If one follows *another* sense of Diderot's *tout*, i.e., Spinoza's God, then Dr Baumann will assuredly deny that the universe is a whole.[90] Maupertuis refuses to believe that there is anything perilous in his system of elementary perceptions, and as such is happy to extend it also to the *largest* bodies in the universe, *without thereby* making gods of them. He criticizes Diderot for believing that he, Maupertuis, is wrong in attributing perception to matter, while on his part attributing *sensation* to matter; with some degree of justification (at least Diderot requires a good deal of interpreting here) he considers this to be a lapse into dualism.[91] There is of course a running ideological thread here, which would be straightforward if it were simply a matter of the two authors trading accusations of atheism, materialism and Spinozism, as they do, but is more complicated because it involves at least three other rhetorical dimensions. Namely, (a) praise for the very same ideas: Diderot describes 'Dr Baumann's' "Erlangen dissertation" as being "full of new and singular ideas," and recommends that "we read his work to learn how to reconcile the boldest philosophical ideas with the deepest respect for religion"[92]; (b) Diderot also suggests that Maupertuis should 'stick to biology' and not address the status of the soul: "if Dr Baumann only kept his system to its rightful limits, and applied his ideas strictly to the

---

89) Ibid., 207-208.

90) Ibid., 208.

91) Ibid., respectively 209, 212. For an attempt to systematize Diderot on this issue (arguing that he is a substance monist but a functional dualist for whom there are levels of complexity in sensitivity), see Jean Deprun, "L'anthropologie de Diderot: monisme métaphysique et dualisme fonctionnel," in A. Mango, ed., *Diderot. Il politico, il filosofo, lo scrittore* (Milan, 1986), 115-122.

92) Diderot, *Pensées*, § L, in *Œuvres complètes*, IX, 79. On the ideological dimension see Aram Vartanian, "Diderot et Maupertuis," *Revue internationale de philosophie* 38:148-149 (1984), 46-66—more as a 'stimulant' than as a guide to Maupertuis or Diderot.

formation of animals..."[93]; (c) lastly, as we have seen, he also deliberately 'overdetermines' what Maupertuis intended.

## 5. Stakes of the Debate

Maupertuis' molecules are a significant 'conceptual persona' in the history of vital *minima*, whether this history is construed as a series of developments and complexifications in matter theory, or a more qualitative, metaphysically laden pre-history of Lavoisian chemistry. They are of course not alone in the relevant natural-philosophical landscape; in addition to Gassendi's *semina rerum* and Glisson's appetitive, perceiving fibres mentioned at the beginning, a more extensive study would have to discuss Buffon's organic molecules, Albrecht von Haller's fibres and the various treatments of molecules in Stahlian and post-Stahlian chemistry, notably Gabriel-François Venel's "molécules intégrantes," which he distinguishes from the 'masses, forces and qualities' which are basic for physics.[94] Admittedly, Venel's chemical definition of living bodies as "mucous bodies" (*corps muqueux*) which undergo "fermentation" itself resolves these bodies into "masses of organic molecules" (*amas de molécules organiques*).[95]

Should we then grant particular significance to the idea of attribution or endowment of properties? What exactly is the nature of the material substrate of the attributes of desire, memory, aversion, intelligence and instinct? It is not obvious exactly at which level Maupertuis wishes to articulate his theory: is it a system of generation among others, or a 'meta-theory' of generation? He clearly wishes to express these ideas at a greater level of abstraction than some others, while however invoking empirical data. But he also is more ambitious than he needs to be, as Diderot will put it, and seeks to provide a metaphysical mooring-point for his theory of generation. Diderot appears to be the more methodologically sophisticated of

93) Diderot, *Pensées*, § LI, in *Œuvres complètes*, IX, 83.

94) Venel, "Chymie," *Encyclopédie*, III (1753), 411b, 416a. On chemistry in the mid-eighteenth century in France, see Kubbinga, *Histoire*, 440-441.

95) Venel, "Muqueux, corps (chymie)," *Encyclopédie*, X (1765), 865b.

the two, since rather than attributing a host of properties to the element (molecule) directly, he recognizes that the sensitivity of an organic being must be explained in terms of its nervous system and the way in which it relates to the individuality or selfhood of this being. In addition, he does not "endow" matter with what Maupertuis himself called "non-physical" attributes[96]—that is, phenomenological properties of matter.

Maupertuis' endowed molecules ultimately tend towards a homuncular theory of an infinitely small entity which possesses (even if only by analogy) the attributes we are familiar with from our own experience. Each molecule possesses a "self" and memories of this "self" which gradually fade away as the molecule enters into increasingly complex arrangements, in a totality, the sense of self of which is stronger than that of the element. Philosophically, this seems open to the charge of pan-psychism (as stated at least as early as Janet's essay in the late nineteenth century). But if understood in heuristic terms, the idea of attributing informational properties to organic, organized matter is ironically more 'realistic' than Diderot's vast organic cosmogony. Conversely, the idea that the properties of complex living beings can only be attributed to the whole, the *organisation*, the network, seems much more convincing to us than Leibnizian, homuncular argumentation (however naturalized). Emergentist theories, to use contemporary terminology,[97] have the advantage that they take into account the properties of the constituent parts, including in a reductionist sense, but then seek to study the consequences of their interaction.

---

96) *Système*, § XXVIII, 155.

97) For a brief application of the term 'emergentism' to Maupertuis, see T.S. Hall, *Ideas of Life and Matter. Studies in the History of General Physiology*, vol. 2 (Chicago, 1969), 26-28. For a more extensive account of eighteenth-century models of organic life as 'organizational' (and thus beyond the split between reduction and emergence), see Wolfe and Terada, "The Animal Economy," 558-574.

# Omnis Fibra Ex Fibra: Fibre Œconomies in Bonnet's and Diderot's Models of Organic Order

**Tobias Cheung**
*Max-Planck-Institute for the History of Science**

**Abstract**
In a long-term transformation, that begins in Antiquity but takes a crucial turn in the Renaissance anatomies, the "fibre" becomes from around 1750 the operative building block and at the same time the first unifying principle of function-structure-complexes of organic bodies. It occupies the role that the cell takes up in the cell œconomies of the second third of the nineteenth century. In this paper, I will first discuss some key notions, technical analogies, and images that are related to "fibre"-concepts from Andreas Vesalius to Albrecht von Haller and then focus on Charles Bonnet's and Denis Diderot's fibre œconomies. In Bonnet's and Diderot's fibre œconomies, the self-active, regulating properties of fibre-agents and their material structures, that reach from fibre bundles, tissues and membranes to apparati of organs, are united within the concrete whole of individual organized "systems" or "networks."

**Keywords**
fibre, system, network, sensation, sensibility, organic, body, organization, œconomy

> All created substances are organized or without organization. The first ones are composed of fibres, possess live, and are called animals or plants. The last ones are massive, have only particles that are put against each other, and are inert.
> – Antoine Le Camus, *Médecine de l'esprit*, 1769

After Giorgio Baglivi's pathology of fibre types and before Xavier Bichat's catalogue of tissue types, Bonnet's and Diderot's main

* Max-Planck-Institute for the History of Science, Boltzmannstraße 22, 14195 Berlin, Germany (tcheung@mpiwg-berlin.mpg.de).

objective is to explain the operative mode and the development of the entire living body through its fibre œconomy. In their anatomical and physiological writings, nearly every organic part is made of fibres and operates through fibres. These fibres can be hard or soft, straight or bent, isolated or interwoven, and hollow or solid; they form tissues, membranes, organs, vessels, apparati of organs, and entire living beings; and they can move themselves, sense, regulate, regenerate, contract into a germ, dilate into web-like structures, and transport data through vibrations or liquids.

From around 1750, the fibre thus becomes, through its passive and active properties, the main operative building block and at the same time the first unifying principle of function-structure-complexes of organic bodies.[1] It occupies the role that the cell takes up in the cell œconomies of the second third of the nineteenth century.[2] The bodies of plants, animals and humans are all constituted through fibres,[3] and their differences mainly rely on different fibre

---

1) Cf. Antoine Le Camus, *Médecine de l'esprit. Où l'on traite des Dispositions & des Causes Physiques qui, en consequence de l'union de l'ame avec le corps, influent sur les opérations de l'esprit; & des moyens de maintenir ces opérations dans un bon état, ou de les corriger lorsqu'elles sont viciées.* ([1st ed. 1753], 2nd. ed. Paris, 1769), 8: "Toutes les substances crées sont organisées ou sans organisation. Les premieres sont composées de fibres, jouissent de la vie, et sont connues sous les noms d'animaux et de végétaux. Les dernieres sont massives, n'ont que des particules appliquées les unes contre les autres et sont inertes. Elles constituent le regne minéral."

2) Cf. Alexander Berg, "Die Lehre von der Faser als Form- und Funktionselement des Organismus. Die Geschichte des biologisch-medizinischen Grundproblems vom kleinsten Bauelement des Körpers bis zur Begründung der Zellenlehre," *Virchows Archiv für Physiologische Anatomie und Physiologie und für klinische Medizin*, 309 (1942), 333-460, 336: "Die Lehre von der Faser nahm in der Naturwissenschaft und Medizin vor Begründung der Zellenlehre eine ähnlich grundlegende Stellung ein wie später diese." Also Hisao Ishizuka, "Enlightening the Fibre-Woven Body: William Blake and Eighteenth-Century Fibre Medicine," *Literature and Medicine*, 25 (2006), 72-92, 72: "... the heart of eighteenth-century medicine was the fibre: it ontologically constituted the animal body, and epistemologically as well as ontologically mediated mind and body; it was the physiological, pathological, and therapeutic underpinning of medical theories of the human body; and it served to substantiate the medicocultural concept of nervous sensibility." See also Paul Ilie, *The Age of Minerva*, 2 vols. (Philadelphia, 1995), vol. 2, 163-202.

3) Fibres can be themselves composed of smaller organic units, as Buffon's organic molecules.

types, structures and properties—plants feed through fibres, animals move and sense through fibres, and humans think through fibres.[4]

Debates about agency, regulation and fibre œconomies shape the concept of 'organism' as an autonomous, self-organizing and self-reproducing entity. They are related to the discursive field of automata as certain kinds of machines that resemble living organic bodies. This field emerges in the Renaissance period together with Aristotelian concepts of self-moving soul-agents.[5] I will point to a specific constellation between weaving automata, their products and the order of the living that appears in the first decades of the eighteenth century. Through this constellation, the organic body can be modeled as an animal, a human or a machine.

Bonnet's and Diderot's fibre œconomies are part of a long-term transformation of different fibre concepts. I will discuss some key notions, technical analogies, and pictorial representations from Andreas Vesalius to Albrecht von Haller. However, I will not refer to the minutiae of the historical development of fibre concepts from Francis Glisson to Baglivi and from Bichat to cell theories.[6] Rather, I give a detailed account of Bonnet's and Diderot's models of organic order in which culminate and intensify transformations of different fibre concepts and their representational modes. This common aspect of their work has rarely been discussed in literature. In *Die Lehre von der Faser als Form- und Funktionselement des Organismus* (1942)—until now the most detailed historical reconstruction of fibre

---

4) For Bonnet, it is still an open question if plants can sense or not. Cf. Charles Bonnet, *Essai analytique sur les facultés de l'âme* (1st. ed. 1760, reprint Hildesheim-New York, 1973), 16.

5) See Silvio A. Bedini, "The Role of Automata in the History of Technology," *Technology and Culture*, 5 (1964), 24-42; Horst Bredekamp, *Antikensehnsucht und Maschinenglauben. Die Geschichte der Kunstkammer und die Zukunft der Kunstgeschichte* (Berlin, 1993); Jessica Riskin, *Science in the Age of Sensibility: The Sentimental Empiricists of the French Enlightenment* (Chicago, 2002).

6) For more detailed studies of these periods, see Berg, "Die Lehre von der Faser"; Mirko Grmek, "La notion de la Fibre vivante chez les Médecins de l'Ecole iatrophysique," *Clio Medica*, 5 (1970), 297-318; François Duchesenau, *La physiologie des lumières. Empirisme, Modèles et Théories* (The Hague, 1982) and *Genèse de la théorie cellulaire* (Montréal, 1987); and Ishizuka, "Enlightening." Ishizuka (ibid., 74) also refers to the model of fibre bodies in the poetry of Richard Blackmore.

concepts—, Berg does not mention Bonnet and Diderot. He is mainly interested in theories of the order and properties of the fibre itself, and not of the entire organic body in which it operates. This lacuna is crucial because in the fibre œconomies of Bonnet and Diderot, in which the fibre is both the most general and the most differentiated building-block of organic bodies, the paradigm of the fibre clearly becomes a unifying and order-generating principle that distinguishes living organic bodies from all other bodies.

For Bonnet and Diderot, organic bodies develop, subsist, and reproduce as if they were automata that produce cloth-like tissues. Comparisons between man-made woven textures and the structure of organic bodies date back to mythologies and natural histories of the Ancient Greeks.[7] However, it is only from the sixteenth century onward that anatomists like Vesalius and Girolamo Fabrizio d'Acquapendente establish a tradition of canal and fibre œconomies in addition to the Hippocratic school that focuses on humoral dispositions to characterize the order of organic bodies (the *eucrasia* or *dyscrasia* of their mixture of fluids).[8] In the second half of the seventeenth century, microscopists like Marcello Malpighi, Antoni van Leeuwenhoek, and Nehemiah Grew initiate a new field of detailed pictorial representations of fibre networks and their micro-structures. Besides the passive properties of these structures (stability, tube-like forms for canal systems, contraction through the influx of liquids), anatomists like Glisson, Baglivi, and Haller assume—within the framework of different ontologies and epistemologies—a specific force (a *vis insita*) that inheres in fibres or certain types of fibres as a cause of their self-moving properties (for example perception, contraction, or irritation). In the second half of the eighteenth century, the discourse about the fibre as the smallest organic building-block and its self-moving properties merges with debates about (1) the development of point-like germs into web-like tissues, tubes and organs, (2) self-regulating processes of organized bodies,

---

7) Cf. Ioanna Papadopoulou-Belmehdi, *L'art de Pandora. La mythologie du tissage en Grèce ancienne*, Thèse de doctorat, École des Hautes Études en Sciences Sociales (Paris, 1992).

8) Cf. Marielene Putscher, "Andreas Vesaliusius (1514-1564)," in *Klassiker der Medizin*, ed. D. v. Engelhardt & F. Hartmann, 2 vols. (Munich, 1991), vol. 1, 116-117.

(3) analogies between organic bodies and tissue producing automata, and (4) the soul-body-relation as a spider-net-, weaver-loom-, puppeteer-marionette or organist-organ-interface.

The first section of the essay refers to continuities and transformations of pictorial representations and explanatory schemes of canal systems and fibre œconomies from Vesalius to Boerhaave and Haller. The second and third section focus on the role of the fibre in Bonnet's and Diderot's models of organic order.

## 1. Canal Systems and Fibre Œconomies from Vesalius to Boerhaave and Haller

In a period of growth and rapid technical development of Europe's weaving industry, a new anatomy of organic bodies takes shape. Anatomists and surgeons as Andreas Vesalius, Jean Fernel, Jean Riolan, Gabriele Fallopio, and Fabrizio compare the bodies of humans and various animals. Their growing critique of Hippocratic humoral physiologies is based on a detailed knowledge of the architecture of solid structures as bones, tendons, and veins. At the end of the sixteenth century, the smallest organic unit that constitutes these structures becomes the "fibre" and its "network."[9]

Web-like representations of canal systems that envelop, penetrate and connect organs and muscles are characteristic for the wood cut illustrations of the various branches of veins in Vesalius's *De humani corporis fabrica* of 1543 (see fig. 1).

For Vesalius, the tissue walls of the veins are formed by a network of straight, oblique and transverse "fibres." Vesalius thus visualizes the three main fibre types that Galen already distinguished in *On the Natural Faculties* (see figure 2).[10]

In Fernel's *Universa medicina*, first published in 1567, "fibres" constitute not only canal systems, but also the solid structure of

---

[9] For fibre and tissue theories in Antiquity, especially in the writings of Greek physicians and philosophers as Erasistratos, Aristoteles, and Theophrast, see Berg, "Die Lehre von der Faser," 337-348.

[10] Cf. Galen, *On the Natural Faculties*, translated by A. J. Brook. (1st ed. 1916, reprint Cambridge, Mass., 1991), 263-281 (Book III, chapters 8-11).

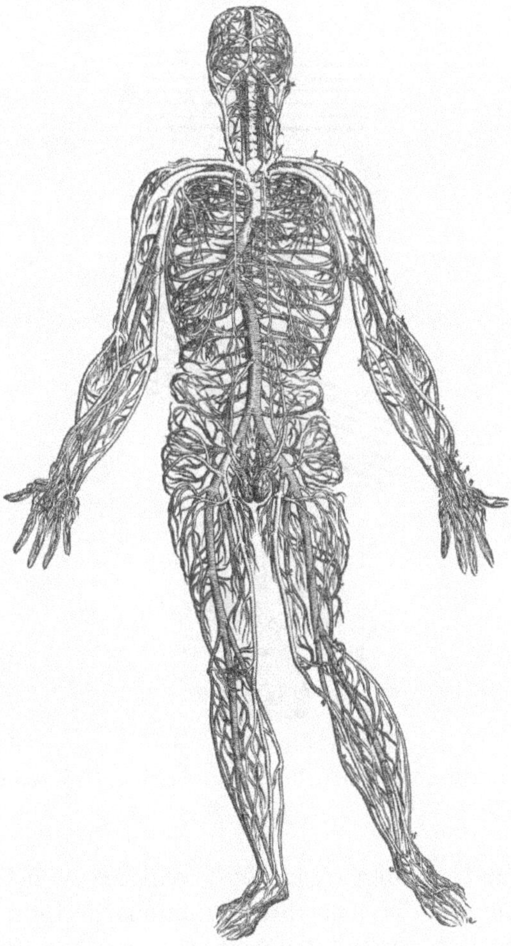

Figure 1: Web of veins. Andreas Vesalius, *De humani corporis fabrica* (Basel, 1543), 450.

the entire body. The "fibre" thus becomes the main structural and operational element of organic bodies.[11] Fernel explains the "nourishment and maintenance" of organic bodies through four "natural faculties" (*facultes naturales*): the "attracting" (*attrahens*), the "expelling"

---

11) Berg, "Die Lehre von der Faser," 352-355, refers to a similar role of the fibre in Fallopio's anatomy.

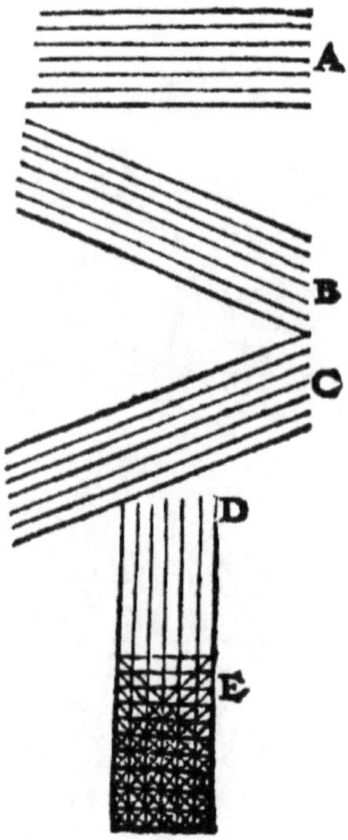


Figure 2: Fibre structure of venous walls. Andreas Vesalius, *De humani corporis fabrica* (Basel, 1543), 437.

(*expellens*), the "retaining" (*continens*), and the "concocting" (*concoquens*) faculty.[12] The first three functions perform mechanical operations through certain structural changes that result into the "assimilation" (*assimilatio*) of nutritive parts to the organization of the body, while the "concoction" transforms the newly integrated parts into the specific mixture of the bodily fluids.[13] In every "organ" or "natural instrument" (*instrumentum naturale*), the attracting faculty depends on the movement of "straight fibres" (*villi recti*), the expelling faculty on "transverse fibres" (*villi transversi*), and the retaining

[12] Jean Fernel, *The Physiologia of Jean Fernel* (1st ed. 1567), translated and annotated by John M. Forrester (Philadelphia, 2003), 321.
[13] Ibid., 323.

faculty on "oblique fibres" (*villi obliqui*).[14] For an experienced anatomist, the "type" and the "variety of fibres" (*genus et varietas villorum*) indicate the different faculties and functions of organic bodies:

> It follows that the number of faculties residing in each part can be discovered from the variety of its functions, and accounts for the different sorts of its fibres. And conversely, the variety of fibres determined by dissection declares the number of faculties of any part: esophagus, stomach, intestines, arteries, veins, womb, both bladders [the gall and urinary bladder], and any general instruments of nature there may be in addition to these.[15]

In *De musculis*, first posthumously published in 1687, Fabrizio combined the fibre œconomy of organic bodies with the process of weaving. He refers to a technical model of the production of the tissue-like "texture" (*textura*) of "muscles," in which the ambivalent term of an organizing "nature" (*natura*), that resides within a fibre complex, replaces the human as a weaving agent. For Fabrizio, the organizing "nature" is itself able "first to dispose fleshy fibres, lengthways produced, like a stretched vertical thread that is in the vernacular [Italian] called 'warp' (*ordimento*), and then to horizontally add flesh as a sub-layer or intersecting fibres that in the vernacular is called 'weft' (*trama*) in a tissue".[16] The "flesh of muscles" thus

---

14) Ibid., 329. In *On the Natural Faculties*, Galen already argued that the movements of organs depend on the setting of their fibres. Cf. Galen, *Natural Faculties*, 263 (Book III, chapter VIII).

15) Fernel, *The Physiologia*, 329. " … Quapropter facultatem parti unicuique assidentium numerus ex functionum varietate cognitus, villorum differentias explicat: econtrario quoque villorum varietas consectione cognita renuntiat quot cuiusque partis sunt facultates, gulae, ventriculi, intestinorum, arteriarum, venarum, uteri, utriusque vesicae, et aliorum si plura his publica naturae sunt instrumenta." However, for Helkiah Crooke in his *Microcosmographia* (1615), the "vessel" is the essential part of the body through which flows "Blood, Heate, Spirits, Life, Motion and Sense." Cf. Gail K. Paster, "Nervous Tension. Networks of Blood and Spirit in the Early Modern Body," in *The Body in Parts. Fantasies of Corporeality in Early Modern Europe*, ed. David Hillman & Carla Mazzio (New York, 1997), 112.

16) Fabrizio Girolamo, *Opera omnia anatomica et physiologica* ([1st ed. 1687], new edition, Leiden, 1738), 392: „Sed non substantiae solum proprietate [the faculty of contraction], sed alia quoque ratione caro musculorum ab omnia alia carne diversa est: quoniam aliarum partium caro uno quidem modo apponitur tanquam intentum stamen, & subtegumen in telis: ut in osse, in cujus constitutione ita natura operatur,

can not only "contract" itself, but also weave organic tissues. Fabrizio calls the tissue-like structure a "membrane" (*membrana*).[17] "Membranes" shape and stabilize most of the organic parts, as bones, inner organs, arteries, and veins.[18] They also serve as the basis for their specific operations.

In the second half of the seventeenth century, the usage and diffusion of microscopic apparati and new injection techniques intensify and differentiate the converging movements between different representational modes of web-like canal systems, fibre œconomies, and functional properties of organic parts. In the last third of the seventeenth century, Giovanni Alfonso Borelli and Claude Perrault's models of organic bodies are characteristic of a shift of interest from the soul-body-interface to the mechanisms of fibre movements.[19]

---

ac si telam esset effectura: primo enim fibras carneas tamquam intentum stamen per longitudinem productas disponit, quod vulgo appellatur l'ordimento: deinde per transversum tamquam subtegmen carnem apponit, fibras scilicet transversas, quae in tela vulgo trama dicitur." Eyvind Bastholm (*The History of Muscle Physiology. From the Natural Philosophers to Albrecht von Haller. A Study of the History of Medicine with a Summary in Danish* [Kopenhagen, 1950], 126) highlights that Girolamo could not see the striation of muscle without a microscope (which he did not use). He suggests that Girolama described membranous structures. See also Berg, "Die Lehre von der Faser," 356.

17) Cf. Girolamo, *Opera omnia*, 392: "Quod clare in infantium, & abortivorum cranii ossibus parumper decoctis intueberis: simili modo. & membranae carnem habent: immo ossa in primo ortu membranae sunt, nequaquam ossa: Quod si membranae, ergo ventriculus, intestina, pericardios, cerebri membranae, venae & arteriae, & aliae universae eodem modo carnem habebunt: eodem modo caro in cerebro, & plerisque omnibus partibus ponitur."

18) Girolamo refers to the ossification of the crane. Cf. Girolamo, *Opera omnia*, 392.

19) Cf. Matthias Tripp, "Le modèle méchanique comme paradigme epsitémologique de la nature et de la pensée aux XVII<sup>e</sup> et XVIII<sup>e</sup> siècles," in *Epistémologie et matérialisme*, ed. O. Bloch (Paris, 1986), 45: "Pourtant il faut se rendre compte du changement - inauguré par Borelli - dans la discussion scientifique concernant la correspondance entre l'âme et le corps. Non seulement l'application stricte de la géometrie au corps vivant, mais aussi le développement et la différenciation de l'analyse fournirent les moyens de restreindre de plus en plus la fonction de l'âme traditionnelle et de favoriser l'extension de la pensee mécanique, meme sous la forme de la machine." Also Dennis Des Chene, "Mechanisms of Life in the Seventeenth Century: Borelli, Perrault, Régis," *Studies in History and Philosophy of Biological and Biomedical Sciences*, 36 (2005), 245-260, 255: "As with Borelli, we see [in Perrault's model of organic order] a definite demarcation

After Robert Hooke's and Niels Stensen's work on different types and mechanical dispositions of natural and artificial fibres, Antoni van Leeuwenhoek distinguishes in *Anatomia seu interiora rerum* (1687) between "bundles of fibres" (*striae*) that are observable with the naked eye and "fibres" (*fibrae*) that can only be observed with the help of lenses.[20] In his *Arcana naturae detecta* (1695), the micro-structure of muscles appears as a web of interwoven fibres (see fig. 3).

While Italian iatromechanists such as Giovanni Alfonso Borelli and Marcello Malpighi refer to the fibre œconomy of animal bodies within the framework of Cartesian micro-machine models and their hydraulic mechanics[21], the anatomist Nehemiah Grew focuses

between the 'mechanics' of animals and what we might call their 'energetics'. The source of active power is the soul, about which very little is said. What remains is to explain the transmission and application of that power to the end of locomotion, and that is a matter of applying mechanical knowledge. Perrault acknowledges that the animal-machine resembles 'pure machines' in the manner of its operation. But every machine requires a mover, and this the mechanism itself is incapable of supplying.".

20) Cf. Antoni van Leeuwenhoek, *Arcana Naturæ, Ope & beneficio exquisitissimorum Microscopiorum. Detecta, variisque experimentis demonstrata* ... ([1st ed. 1687] 3rd ed. Leiden, 1808 [1708]), I, 49; and Berg, "Die Lehre von der Faser," 383-384. For the usage of microscopes in the seventeeenth century, see Luigi Belloni, "Marcello Malpighi and the Founding of Anatomical Microscopy," in *Reason, Experiment, and Mysticism*, ed. M. L. R. Bonellli & W. R. Shea (New York, 1975), 95-110; Brian J. Ford, *Single Lens. The Story of the Simple Microscope* (Heinemann, 1985); Catherine Wilson, *The Invisible World. Early Modern Philosophy and the Invention of the Microscope* (Princeton (New Jersey), 1995); and David Freedberg, *The Eye of the Lynx: Galileo, his Friends, and the Beginnings of Modern Natural History* (Chicago et al., 2002).

21) Cf. Marcello Malpighi, *Opera Omnia. Zwei Bände in einem Band* (1st ed. 1686, reprint Hildesheim, 1975), vol. 2, 128 ('De Polypo Cordis'); Berg, "Die Lehre von der Faser," 337-348; and Domenico Bertoloni Meli, "Blood, Monsters, and Necessity in Malpighi's De Polypo Cordis," *Medical History*, 45 (2001), 511-522, 518-519. For Malpighi's micro-machine-models, see Marcello Malpighi, *Marcelli Malpighii Opera posthuma figura aenis illustrata quibus praefixa eiusdem vita a seipso scripta* (London, 1697), 110-111; Howard B. Adelmann, *Marcello Malpighi and the Evolution of Embryology*, 5 vols. (Ithaca, 1966), vol. 1, 570; Wilson, *Invisible World*, 232-236; and François Duchesneau, "Malpighi, Descartes, and the Epistemological Problems of Iatromechanism," in *Reason, Experiment, and Mysticism*, ed. M. L. R. Bonellli & W. R. Shea (New York, 1975), 111-130, and *Les modèles du vivant de Descartes à Leibniz* (Paris, 1998), 196-208. Berg, "Die Lehre von der Faser," 356-358 discusses Descartes's fibre theory.

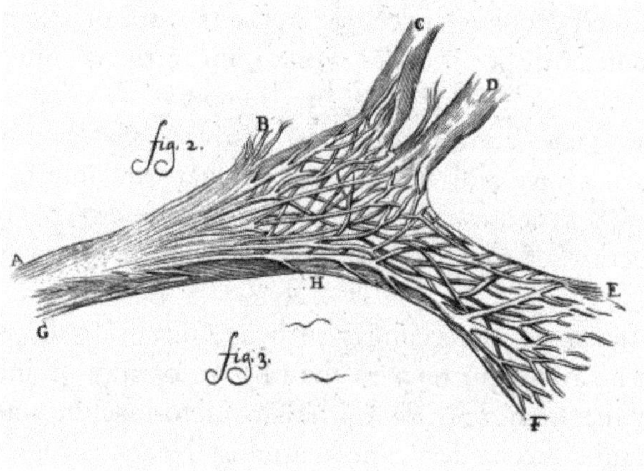


Figure 3: Microstructure of a muscle fibre (Leeuwenhoek 1695). Published in Antoni van Leeuwenhoek, *Opera Omnia*, vol. 1 (Leiden, 1715), 46.

on the tissue-structure of plants. In *The Anatomy of the Trunks*, a text that was first presented to the Royal Society in 1673 and published in *The Anatomy of Plants* (1682), Grew compares the "tissue" of plants with the "bone-lace" that is produced for cushions:

> So that the most unfeigned and proper resemblance we can at present, make of the whole Body of a Plant, is, to a piece of fine Bone-Lace, when the Women are working it upon the Cushion. For the Pith, Insertions, and Parenchyma of the Barque, are all extream Fine and Perfect Lace-Work: the Fibres of the Pith [Mark] running Horizontally, as do the Threds in a Piece of Lace; and bounding the several Bladders of the Pith and Barque, as the Threds do the several Holes of the Lace; and making up the Insertions without Bladders, or with very small ones, as the same Threds likewise do the close Parts of the Lace, which they call the Cloth-Work. And lastly, both the Lignous and Aer-Vessels, stand all Perpendicular, and so cross to the Horizontal Fibres of all the said Parenchymous Parts; even as in a Piece of Lace upon the Cushion, the Pins do to the Threds. The Pins being also conceived to by Tubular, and prolonged to any length; and the same Lace-Work to be wrought many Thousands of times over and over again, to any thickness or hight according to the hight of any Plant. And this is the true Texture of a Plant: and the general composure, not only of a Branch, but of all other Parts from the Seed to the Seed.[22]

---

22) Nehemiah Grew, *The Anatomy of Plants* (1st ed. 1682, reprint New York, 1965), 121-122.

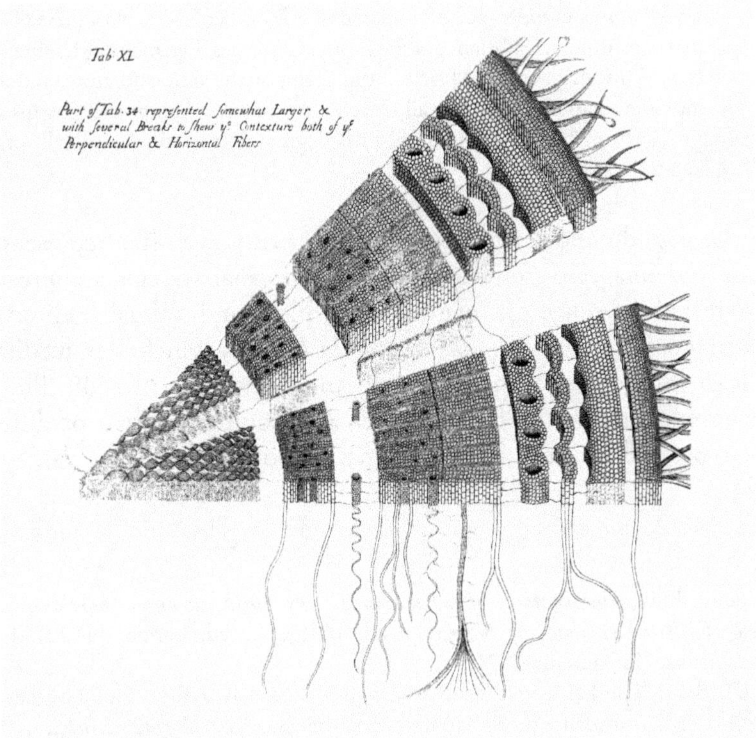


Figure 4: Microstructure of plant tissue. Nehemiah Grew, *The Anatomy of Plants* (London, 1682), Plate 40.

Grew's illustrations of the "texture" of plants thus look like a woven piece of bone-lace (see fig. 4).

In *The Anatomy of the Humane Body Abridged*, first published in 1698 with a seventh edition in 1723, Grew's colleague James Keill transformed the human body into a network of different fibre types:

> ... all the Parts are made up of Threads, or Fibres, of which there be different Kinds; for there are some soft, flexible, and a little elastick; and these are either hollow, like small Pipes, or spongeous, and full of little Celles, as the nervous and fleshy Fibres; others there are more flexible, but with a strong Elasticity or Spring, as the membraneous and cartilaginous Fibres; and a third Sort are hard and inflexible, as the Fibres of the Bones. Now of all these, some are very sensible, and others are destitute of all Sense; some so very small as not to be easily perceived; and others, on the contrary, so big as to be plainly seen. And most of them, when examined

> with a Microscope, appear to be composed of still smaller Fibres. Now these Fibres do first constitute the Substance of the Bones, Cartilages, Ligaments, Membranes, Nerves, Veins, Arteries and Muscles. And again, by the various Texture, and different Combination of some or all these Parts, the more compound Organs are framed; such as the Lungs, Stomach, Liver, legs, and Arms, the Sum of all which make up the Body.[23]

In the first decades of the eighteenth century, web-like representations and images of fibre œconomies and canal systems are already a standard reference for anatomists of plant and animal bodies. In his manual *Corporis humani anatomia* (1693), which was reedited at least thirteen times in both Latin and German until 1739, Philip Verheyen portrays the human body as a complex system of different types, dispositions and combinations of fibres (see fig. 5).[24]

---

23) James Keill, *The Anatomy of the Humane Body Abridged; or, A Short and Full View of All the Parts of the Body* ([1st ed. 1698], 7th ed. London, 1723), 1-2. Cf. Ishizuka, "Enlightening," 74.

24) Cf. Berg, "Die Lehre von der Faser," 390-392; and Rau Suy, "Philip Verheyen (1648-1710) and his Corporis Humani Anatomiae," *Acta Chirurgica Belgica*, 107 (2007), 343-54. Boerhaave combines the notion of a hollow "nerve fibre" as the basic building-block of all organic structures with Frederik Ruysch's visualization of canal systems through various injection techniques. In Boerhaave's hydraulic model of organic bodies, "canals" or "tubes" (*canales*) are differentiated into smaller tubes that "finally form, branched within themselves, the structure of a net (*retis structura*)." Cf. Hermann Boerhaave, *De usu ratiocinii Mechanici in Medicina. Oratio habita in auditorio magno. 24. September 1703* (Leiden, 1703), 10: "… canalis est conicus, elasticus, inflexus, divisus in similes minores eodem trunco ortos, qui ultimo … retis structura in se mutuo patent." In *Vernünfftige Gedancken. Von dem Gebrauche der Theile im Menschen, Thieren und Pflantzen* (1725), Christian Wolff also refers to the web-structure of organic bodies. Cf. Christian Wolff, *Vernünfftige Gedancken. Von dem Gebrauche der Theile im Menschen, Thieren und Pflantzen*, edited by J. École et al., (1st ed. 1725, reprint Hildesheim, 1980), 481-482 (§ 174): "Denn ob gleich der gantze Leib nichts anders als ein Gewebe von kleinen Röhrleinen ist / die von den grossen wie die kleinen Fäserlein in den Wurtzeln von ihren grösseren Theilen abstammen; so finden wir doch nirgends / daß die Theile in ihrem Anfange von andern abgesondert darliegen; vielmehr ist alles bis auf das kleineste mit einander verbunden …" For Wolff's model of organic order, see Tobias Cheung, "Ontologie und Physiologie bei Christian Wolff (1679-1754)," *Verhandlungen zur Geschichte und Theorie der Biologie*, 10 (2004), 263-281.

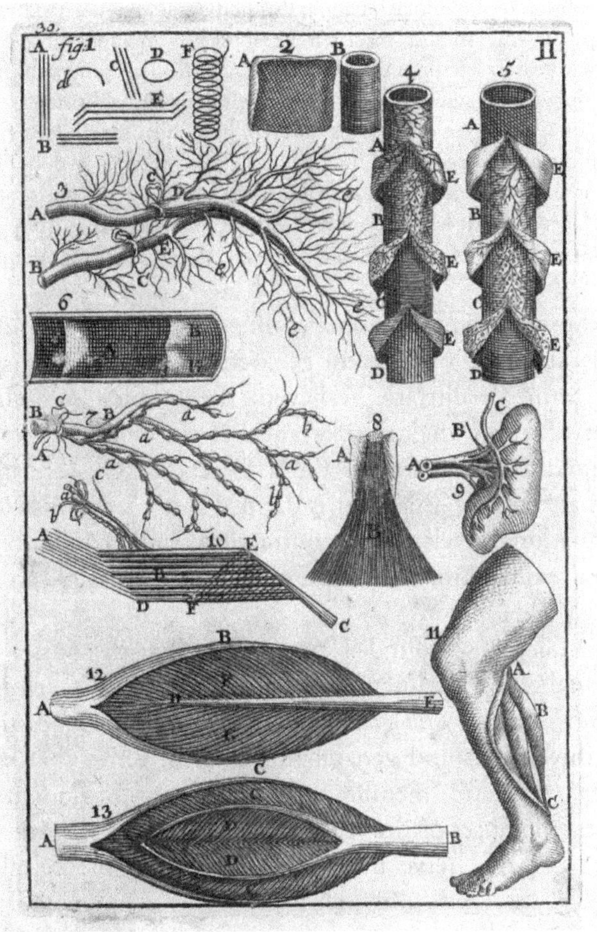

Figure 5: Fibre œconomy of the human body. Philip Verheyen, *Corporis humani anatomiæ*, 2 vols. (New edition Cologne, 1712-1713) vol. 1 (1712), 30, Tab. II.

In *Oeconomia regni animalis* (1740-41), Emanuel Swedenborg finally reduced the different fibre types to a "simple fibre" out of which all other non-fluid "continuous" organic material[25] and the soul-body-interface is formed:

---

25) Emanuel Swedenborg, *The Economy of the Animal Kingdom considered anatomically, physiologically and philosophically*, translated from the Latin and edited by A. Acton (Philadelphia, 1918), 163 (§ 250): "I make a distinction between continuous and con-

> In the subjects of the animal kingdom this simplest fibril is deservedly to be called the first, supreme, inmost, most remote, most universal and most perfect fibre, yea, the one only continuous substance from which all other continuous substance, howsoever many they be, derive their existence, subsistence, nature and faculty of action ... So that to this fibre can be applied the philosophical canon: That in the animal system, in that it is an animal, there is nothing that is truly substantial except this fibre, and the rest of the compositions consist of mere accidents ... Moreover it is proper to predicate of the fibre that it is the pathway of the determinations of the soul, and, as it were, the ray of its intellectual light ...[26]

Besides the representation of web-like canal systems and a great variety of catalogues of different geometrical shapes and dispositions of fibres from Vesalius to Verheyen, the problem of self-moving properties of fibres characterizes another set of explanatory strategies within the fibre debates. From Descartes, Glisson and Willis to Baglivi and Haller, this layer is divided into mechanistic or chemical explanations of elasticity, contractility, and irritability and so-called vital explanations of these phenomena through an inherent force.[27]

Haller intervenes around 1750 in the fibre discourse. He calls the human body a "cellular fibre fabric" (*cellulosa fabrica fibris*).[28] In *Primae lineae physiologiae* (1747), Haller clearly distinguishes between three types and general functions of fibres that constitute all "tissues" (*tela*) and "membranes" (*membranae*) of human and animal bodies: the fibres of the "cellular tissue" (*cellulosa tela*), the muscle fibres, and the nerve fibres.[29] The fibres of the cellular tissue

---

tiguous substances; or, what is the same thing, between coherent and fluid substances. For continuous and coherent substances constitute tunics, while contiguous or fluid substance permeate the little canal formed by the tunic."

26) Ibid. (§§ 250-1).

27) Cf. Owsei Temkin, "The Classical Roots of Glisson's Doctrine of Irritation," *Bulletin of the History of Medicine*, 38 (1964), 297-328; Julian Jaynes, "The Problem of Animate Motion in the Seventeenth Century," *Journal of the History of Ideas*, 31 (1970), 219-34; and Ishizuka, "Enlightening."

28) Albrecht von Haller, *Primae lineae physiologiae in usum praelectionum academicarum* (Göttingen, 1747), 7 (§ 17).

29) Cf. ibid., 11 (§ 23): "Summam dignitatem huius cellulosae telae adgnoscet, qui cogitabit, ab ea sola pendere omnium arteriam, nervorum, fibrarum muscularium, adeoque compositarum inde carnum viscerumque legitimam firmitatem, stabilitatem

serve as the basic stabilizing material for organic structures, irritable muscle fibres can contract themselves through a specific *vis insita*, and nerve fibres transmit sense data or instructions from a regulating, soul-like agent.

After Haller, Bonnet and Diderot establish their fibre œconomies. They both continue to work with pictorial representations and explanatory schemes of canal systems and web-like interwoven structures that possess active properties as contractility and irritability.

## 2. Bonnet's Living Looms of Germ-Fibre-Units

After his observations of the parthenogenesis of aphids and Abraham Trembley's experiments on the complete regeneration of artificially detached parts of the fresh water polyp *Hydra* in 1744, Bonnet works on a germ-fibre-theory to explain the development of organic bodies.[30] If pieces of something living can again grow to entire organic bodies with heads and arms or stems and branches, then, Bonnet argues, there have to be small organized units with reproductive potentials within the visible units. These small, point-like

... Ea cum vasi, nervis, fibris muscularibus & tendinieis ... omnia viscera, omnes musculos, glandulasque & ligamenta, & capsulas componit, ab ea sola, eiusque varia longitudine, tensione, copia, proportione, glandularum viscerumque diversitatem oriri, & ab ea denique longe maximam partem corporis ipsius effici certum est, si quidem non totum ex huiusmodi fillis cellulolis componitur."

30) For Bonnet's germ-fibre-theory, see Lorin Anderson, *Charles Bonnet and the Order of the Known* (Dordrecht, 1982); Olivier Rieppel, "The Reception of Leibniz's Philosophy in the Writings of Charles Bonnet (1720-1793)," *Journal of the History of Biology*, 21 (1988), 119-145; Mario Buscaglia, "Bonnet dans l'histoire de la méthode expérimentale," in *Charles Bonnet (1720-1793): Savant et Philosophe. Actes du colloque international de Genève*, ed. M. Buscaglia et al. (Genève, 1994), 283-316; Tobias Cheung, "Die Ordnung des Organischen. Zur Begriffsgeschichte organismischer Einheit bei Charles Bonnet, Spinoza und Leibniz," *Archiv für Begriffsgeschichte*, 46 (2004), 87-108; "Charles Bonnets allgemeine Systemtheorie organismischer Ordnung," *History and Philosophy of the Life Sciences*, 26 (2005), 177-207, and "The Hidden Order of Preformation: Organized Bodies in the Writings of Louis Bourguet, Charles Bonnet and Georges Cuvier," *Early Science and Medicine*, 11 (2005), 11-49; and François Duchesneau, "Charles Bonnet's Neo-Leibnizian Theory of Organic Bodies," in *The Problem of Animal Generation in Early Modern Philosophy*, ed. J.E.H. Smith (Cambridge, 2006), 283-316.

units are germs that extend their contracted fibre œconomy like "mini-machines" (*machinules*) into the "organization" of adult bodies.[31] However, like Leibniz, Bonnet emphasizes that the "system" of these mini-machines cannot be adequately explained by mechanistic theories. Rather, it is evident for Bonnet that the systemic interaction of mini-machines presupposes the existence of a plan-structure-complex that regulates all actions and the development of organic or "organized bodies" (*corps organisés*). The germ-fibre-unit of an individual organized body thus is a preexisting system of parts with basic regulating and structural properties:

> … I understand in general by the word germ all these preordinations or preformations of parts that are themselves able to determine the existence of a Plant or an Animal (*de parties capables par elle-même de déterminer l'existence d'une Plante ou d'un Animal*).[32]

In *Considérations sur les corps organiques* (1762), Bonnet introduces his model of a germ-fibre-unit as an automaton that is composed of indefinitely small fibres. These fibres interact as "mini-machines" (*machinules*) to "animate" the *tout organique*: "A fibre, as simple as it might appear, is nonetheless a *tout organique*, that feeds, growths and vegetates."[33] The *tout organique* can differentiate into nine thresholds of organized units: the germ, the fibril, the fibre, the "mesh" (*maille*) or the "cell," the "fascicle" or "fibre bundle" (*faisceau*), the tissue, the organ, the apparatus, and the entire organism. Each higher organized unit is composed of lower units.

---

31) Cf. Charles Bonnet, *Contemplation de la nature*, in Bonnet, *Œuvres d'Histoire naturelle et de Philosophie*, 10 vols. (Neuchâtel, 1779-1783), vol. 4/1, 360-361. In *De motu animalium* (1680-1), Giovanni Alfonso Borelli often referred to the Latin expression *machinulae* to characterize mini-machines of fibre complexes in organic parts as muscles.

32) Bonnet, *Tableau des Considérations sur les corps organisés*; in Charles Bonnet, *Palingénésie* (1st ed. 1769, reprint Paris, 2002), 82. Cf. Charles Bonnet, *Considérations sur les corps organisés* (1st ed. 1762, reprint of the *Œuvres* edition (1779-1783, vol. 3; Paris, 1985), 63 (§ 83): "L'organisation primitive des germes détermine l' arrangement que les atômes nourriciers doivent recevoir pour devenir parties du tout organique."

33) Bonnet, *Palingénésie*, 257. Bonnet describes sometimes also "fibrils" (*fibrilles*) as "mini-machines." Cf. Bonnet, *Contemplation de la nature*, in Bonnet, *Œuvres*, vol. 4/2, 361; and *Palingénesie*, 232 and 262.

The fibre that is "supposed to be simple is itself composed of a multitude of "fibrils" (*fibrilles*) that are, as for themselves, composed of a multitude of more or less homogeneous particles (*molécules*) as their first elements, while the fibrils are the second elements."[34] Each fibre represents in general a basic fusion point of the external particles with the inner order of the organic body. It possesses assimilative properties: "... the fibre determines through the Mechanics of its Structure the arrangement of its nutritive Atoms ..."[35]

From this fusion result the characteristics of the "primordial layers" of higher organs in form of specific "patterns" (*trames*). These patterns can appear as "meshes" (*mailles*) or "cells." They belong again to various types of "fibre bundles" (*faisceaux*) and "tissues."

Lower units, as for example the parenchymatic or cellular tissue[36], are less determined in their operative effects than higher units (organs or apparati). Cells, bundles of strings and tissues represent the structural or "solid" physiological basis of the organic "network" (*réseau*). The organs and the apparati operate through the co-existing and co-ordinated "meshes" of the "network":

> ... we learn from physiology that there is no organic part that is not covered externally and internally with a cellular or parenchymatic tissue. It is so universally extended that it enfolds the whole system of fibres. One could thus think of it as the principal instrument of growth. The diverse incrustations or incorporations that determine the consistence [of organic parts], the [process of] growth and the most essential modifications of each part are situated within its nearly infinitely varied meshes or pores.[37]

---

34) Ibid., 258.

35) Bonnet, *Essai analytique*, 68 (§ 101). Cf. Bonnet, *Contemplation de la nature*, in Bonnet, *Œuvres*, vol. 4/1, 152: "Mais les fibres sensibles se nourrissent comme toutes les autres parties du corps: elles *s'assimilent* ou s'incorporent les matieres alimentaires; elles croissent, et tandis qu'elles se nourrissent et qu'elles croissent, elles continuent à s'acquitter de leurs fonctions propres; elles demeurent essentiellement ce qu'elles sont. Leur méchanique est donc telle, qu'elles s'incorporent les matieres alimentaires dans un rapport direct à leur structure et à leurs déterminations acquises."

36) Cf. Bonnet, *Palingénésie*, 288.

37) Ibid., 296.

The organic "network" thus is a "networking system" (*ouvrage à réseau*)[38] of organized units on various levels, a system of a "nearly infinite" number of intermediate "chain-links" (*chaînons*) that appear in certain subordinated levels of order. This networking system depends on the "enchained" microscopic structures of "bended," "stretched" and "hooked" fibres or "threads" (*fils*) that form "meshes" and "tissues."

To describe the growth of the organic "network," Bonnet often refers to the process of weaving and to the mechanism of looms.[39] The folded skins of organs become thus different kinds of weaving patterns or "cloths" (*étoffes*):

> The organized bodies are more or less fine tissues, they are networks, kinds of cloths of which the chain forms itself the pattern of an art that we would have to admire even if it would be known to us ...The bodies of plants and of animals are thus sorts of looms, more or less composed machines that convert the diverse materials that are used for the activities of their mechanisms (*ressorts*) and their fluids.[40]

Bonnet's loom analogy reflects the invention of new machines of spinning and weaving processes for the fast growing market of the weaving industry, especially in England and France.[41] It was not before 1785 that Edmond Cartwright patented the first fully mechanized loom.[42] However, the automatization of spinning machines and looms already begun in the first half of the eighteenth century. In 1733, John Kay invented the so-called flying shuttle.[43] The weaver

---

38) Bonnet, *Essai analytique*, 65 (§ 97).

39) Cf. Tobias Cheung, *Res vivens. Agentenmodelle organischer Ordnung 1600-1800* (Freiburg im Breisgau, 2008), 213-229.

40) Bonnet, *Contemplation de la nature*, in Bonnet, *Œuvres*, vol. 4/2, 359-360. Cf. Bonnet, *Considérations*, 63 (§ 83): „Un solide non-organisé est un ouvrage de *marquetterie*, ou de pièces de rapport. Un solide organisé est une étoffe formée de l' entrelacement de différents fils. Les *fibres élémentaires* avec leurs *mailles*, sont la *chaîne* de l'étoffe; les atomes nourriciers qui s'insinuent dans ces mailles, sont la *trame*."

41) Cf. Gustav Baum, *Die Baumwoll-Spinnerei und Weberei in ihrer bautechnischen und maschinellen Entwicklung* (Berlin, 1913), 25-42.

42) Four years later, a steam engine powered the movements of the loom.

43) In the same year, John Wyatt invented an automated spinning machine.

could move Kay's shuttle, itself loaded with cotton, only with a flick of the wrist of one hand from one side of the loom to the other and change at the same time the pattern of the warp threads.[44] In 1745, Jacques Vaucanson developed a completely automated process of the production of woven patterns through punch cards.[45] But Vaucanson's machine remained a prototype until 1805 when Joseph-Marie Jacquard combined Cartwright's power loom with an automated punch card system.

In his *Lettres philosophiques* (1729), Louis Bourguet already compared the mechanism of looms with the mechanism of organic bodies. However, Bourguet focussed on the automated process of weaving itself and not on the fibre œconomy of the product:

> The Mechanic Arts still reveal a couple of other good examples, and there is one among all the others that I have to mention because it seems to be truly accurate for a better understanding of the subject that we examine [i.e. organic bodies]. I would like to talk about the vertical loom[46] that is used to produce *Rubans, Galons*, and all kinds of *Cloths* with gold, silver, silk, wool and linen. This work of art is such that a simple Worker, who understands nothing, can produce all theses nice Works in passing one or a couple of shuttles along the Chain and in putting the feet on a certain number of Pedals.[47]

---

44) Cf. Almut Bohnsack, *Spinnen und Weben. Entwicklung von Technik und Arbeit im Textilgewerbe.* ([1st ed. 1981] 2nd ed., Bramsche, 2002), 154-156.

45) Cf. Alfred Barlow, *The History and Principles of Weaving by Hand and Power* ([1st ed. 1878] 2nd ed. London, 1879), 141: " … M. Bonchon, in 1725, employed a *band of pierced paper* pressed by a hand-bar against a row of horizontal wires, so as to push forward those which happened to lie opposite the blank spaces, and thus bring loops at the lower extremity of vertical wires in connexion with the comb-like rack below, &c. In 1728 M. Falcon substituted a chain of cards, and a square prism (known as the cylinder) in lieu of the band of paper of Bonchon. In 1745, Vaucanson suppressed altogether the cumbrous tail-cards of the draw loom, and made the loom completely self-acting by placing the pierced paper or card upon the surface of a large pierced cylinder, which travelled backwards and forwards at each stroke, and revolved through a small angle by ratchet work."

46) *Haute-Lice.* The threads of this loom are disposed vertically. In the "*basse lice*," they are disposed horizontally.

47) Louis Bourguet, *Lettres philosophiques sur la formation des sels et des crystaux et sur la génération et le méchanisme organique des Plantes et des Animaux* (Amsterdam, 1729), 145.

For Bonnet, it is both the automated production process and the product that are similar in looms and living beings, although the comparison should not be pushed too far.[48] Bonnet explains the order of organic bodies through the assimilative "reproduction" of the order of a regulating "plan" within a "system" of mini-machines that produce, sustain and repair "tissues." These regulated processes are executed by a "secret force" (*force secrète*) that acts like a self-moving flying shuttle. The shuttle "chases the nutrition into the meshes (*mailles*)."[49]

However, in difference to the mechanism of looms, organic systems react to outer "stimuli." The regulating activity of the preexisting "plan" depends on a continuous initiation of movements through various "impulse" series.[50] Such "impulses" emerge in muscles and nerves. While Haller's irritability of isolated muscles served Bonnet as a standard reference for an experimental proof of the general existence of this property in organic units, he tried to combine Haller's irritability with the "impulse" series of nerve fibres to describe a general property of the natural automaton to initiate and repeat coordinated movements. Such a system of impulses and movements is for Bonnet an expression of the "vital principle" (*principe vital*) of living beings.[51]

Bonnet localizes the natural spinning unit of tissues in an "evolving" germ that "contains"[52] ever smaller and smaller germs, an image that is similar to Malebranche's example of an apple tree that is contained in its seeds.[53] But every imaginative approach to the process of *emboîtement* risks to cut short a proper understanding of

---

48) Cf. Bonnet, *Considérations*, 63 (§ 83).

49) Bonnet, *Contemplation de la nature*, in Bonnet, *Œuvres*, vol. 4/1, 259.

50) Cf. Bonnet, *Essai analytique*, 508-9 (§ 797).

51) Cf. Bonnet, *Palingénésie*, 267.

52) Beside of the *emboîtement*-theory of germs, Bonnet also discussed in *Considérations sur les corps organisés* a dissemination-theory. He finally preferred the *emboîtement*-theory. Cf. Raymond Savioz, *La philosophie de Charles Bonnet de Genève* (Paris, 1948), 75-78.

53) Cf. Bonnet, *Mémoires sur les germes*, in Bonnet, *Œuvres*, vol. 5, 10. Savioz, *La philosophie*, 76-77 and Jacques Marx, *Charles Bonnet contre les Lumières 1738-1850*, 2 vols. (Oxford, 1976), 57-58 discuss the influence of Malebranche on Bonnet's preformation theory.

parts of matter that could be indefinitely small, and yet organized.[54] Such an indefinitely small, but organized matter is not simply an "abridged sum" (*abrégé raccourci*) of its "evolved," visible body. Rather, Bonnet emphasizes, it is necessary "to give a much larger signification to the notion of germ."[55]

In his *Principes philosophiques* (1754), Bonnet thought of the preexistence of a similar, though smaller organic body in the germ that "evolves" during its development.[56] The publication of Haller's research on the development of embryos in the egg yolk of chickens in 1757 seemed to confirm Bonnet's viewpoint.[57] But from the *Considérations sur les corps organisés* (1762) to the *Palingénésie philosophique* (1769), Bonnet changes his position towards a more dynamic, regulative process that characterizes organic development:

> When the chicken is still in the state of a *germ*, all its parts have forms, proportions and positions (*situations*) that differ extremely from those in which evolution will cover them. This goes so far that, if we could see this small germ in larger form, it would be impossible for us to recognize it as a chicken ... In brief, the germ is composed of a sequence of points that afterwards form a line. These lines prolong and multiply themselves, and they produce surfaces.[58]

Bonnet's developmental point-line-surface-scheme resembles the geometrical scheme of Boerhaave's pupil Johann Friedrich Schreiber. In

---

54) Cf. Bonnet, *Mémoires sur les germes*, in Bonnet, *Œuvres*, vol. 5, 2.

55) Bonnet, *Palingénésie*, 257.

56) Bonnet, *Principes philosophiques*, in Charles Bonnet, *Essai de psychologie* (1st ed. 1755, reprint Hildesheim, 1978), 210.

57) Cf. Jacques Roger, *Les sciences de la vie dans la pensée française au XVIII^e siècle* ([1st ed. 1963] 2nd ed. Paris, 1993), 724.

58) Bonnet, *Considérations*, 481 (§ 351). Cf. Charles W. Bodemer, "Regeneration and the Decline of Preformationism in Eighteenth Century Embryology," *Bulletin of the History of Medicine*, 38 (1964), 20-31, 25-27. However, in the *Palingénésie*, Bonnet still argues that a similar, though smaller organic body preexists in the germ. Cf. Bonnet, *Palingénésie*, 260: "Mais quand il est question d'expliquer la reproduction d'un tout organique dissimilaire, il me paroît, que je suis dans l'obligation philosophique d'admettre, que ce tout prééxistoit dans un germe proprement dit, où il étoit dessiné très en petit et en entier. J'admets donc, qu'une tête, une queue, une jambe prééxistoient originairement sous la forme de germe, dans le grand tout organique où elles étoient appellées à se développer un jour."

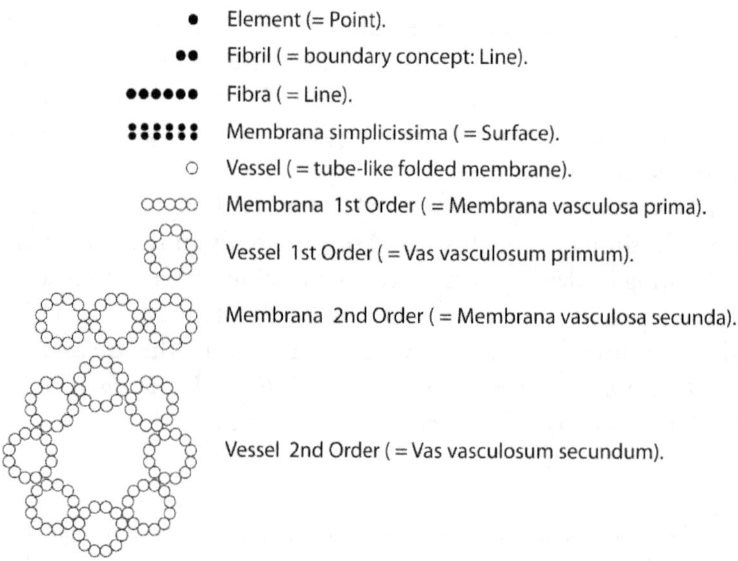


Figure 6: Point-line-surface-scheme. Johann Friedrich Schreiber, *Elementorum medicinae physico-mathematicorum tomus I.* (Frankfurt/Leipzig, 1731), as reconstructed by Alexander Berg, "Die Lehre von der Faser als Form- und Funktionselement des Organismus," 411.

*Elementorum medicinae physico-mathematicorum tomus I.* (1731),[59] Schreiber constructs organic forms out of lines and surfaces that are composed of infinitely small monadic points or atomic "elements." These "elements" are juxtaposed through a cohesive force (see figure 6).[60]

Besides the discussions on developmental processes, Bonnet's model of organic bodies and living looms is also part of debates on the soul-body-interface. In his *Essai de psychologie* (1755), Bonnet combines the image of a soul that plays on the body like an organist on an organ with the fibre fabric of the evolved germ:

---

[59] Christian Wolff wrote the preface to this text.

[60] Berg, "Die Lehre von der Faser," 411. There is no figure in Johann F. Schreiber's *Elementorum medicinae physico-mathematicorum tomus I.* (Frankfurt-Leipzig, 1731); for Schreiber's description of the construction of organic forms, see 175-229.

> The seat of the soul is a small machine ... One can imagine this admirable instrument of the operations of our soul as a clavichord, an organ, a clock, or some other, much more complicated machine ... The soul is the musician that executes different melodies on that machine, or that judges about those that are executed in the machine and that it repeats. Each fibre is a kind of a key or a hammer destined to produce a certain tone. It does not matter if the keys are moved by the objects or by the moving force of the soul; it is always the same play. It can only differ in length or intensity.[61]

From Willis's *Cerebri anatome* (1664) and William Croone's *De ratione motus musculorum* (1664) to Baglivi's *De fibra motrice et morbosa* (1700) and David Hartley's *Observations on Man* (1749), the fibre fabric of organic bodies has been associated with the transmission of stimuli and sensations through vibrations that originate

---

61) Bonnet, *Essai de psychologie*, 13-14. Cf. Bonnet, *Essai analytique*, 15-16 and 52. From Descartes, Pascal and Charles Perrault to Diderot, comparisons between clavichords or organs and the fibre mechanisms of organic order are frequently employed. Diderot compares the living body with a "sensible clavecin" (*clavecin sensible*). Cf. René Descartes, *Traité de l'homme*, in *Œuvres de Descartes*, ed. C. Adam & P. Tannery, 11 vols. ([1st ed. 1897-1913] New Edition, Paris, 1996), vol. 11, 165; Charles Perrault, *Essais de physique ou recueil de plusieurs traitez touchant les choses naturelles*, 4 vols. (Paris, 1680-88), vol. 3 (1680), Avertissement, 1; and Denis Diderot, *Entretien entre d'Alembert et Diderot. Le rêve de d'Alembert. Suite de l'entretien* (Paris, 1965), 50-51. For further references, see Jacques Proust, "Variations sur un thème de l'«Entrétien avec D'Alembert»," *Revue des sciences humaines*, 112 (1963), 435-454; Jamie C. Kassler, *Inner Music. Hobbes, Hooke, and North on Internal Character* (London, 1995), 43-48; and Jean Mayer, *Diderot. Homme de science* (Rennes, 1959), 269-70 and 320-2. Another model of the soul-body-interface is the puppeteer-marionette-relation. La Mettrie mentions this interface in *L'Homme-machine* (1748), and Leonhard Euler refers to it in the first volume of *Briefe an eine deutsche Prinzessin* (1769). Cf. Julien-Offray de La Mettrie, *Œuvres philosophiques*, texte revu par Francine Markovits, 2 vols., (1st ed. 1751, reprint Paris, 1987), vol. 1, 103; and Leonhard Euler, *Briefe an eine deutsche Prinzessin über verschiedene Gegenstände aus der Physik und Philosophie*, 3 vols. (1st ed. 1769-1773, 2nd ed.) (Leipzig, 1773-1774), vol. 2 (1773), 9: "Was die ... Verbindung betrifft, durch welche die Seele, indem sie auf die äußersten Spitzen der Nerven wirkt, nach ihrem Gefallen die Glieder des Körpers bewegen kann; so mögte man sie mit einem Marionetten-Spieler vergleichen, der nur einen gewissen Faden ziehn darf, um die Marionetten nach seinem Gefallen herumgehen und ihre Glieder bewegen zu lassen. Dieses Gleichnis ist aber nur sehr unvollkommen. und das Band zwischen Seele und Körper ist unendlich genauer." (Thanks to Hanns-Peter Neumann for this reference).

in a soul-agent and in the senses.[62] However, in his later texts, Bonnet rejects—like Louis de Lacaze in his *Idée de l'homme physique et moral* (1755)—a mechanical, string-based model for the soul-body-interface. In the nerve system, the impulse series seems for Bonnet to depend on an "electric fluid," instead of being the result of minimal "vibrations":

> The nerves that have been illustrated as strings of a music instrument are not at all under tension like the strings of such an instrument. They are not, like them, made to oscillate; they are not, like them, stretched in a straight line; they appear to have a lot of inflections. Finally, there are neither elastic nor irritable. Their proper substance is malleable and mushy ...[63]

---

62) Cf. Robert B. Glassman & Hugh W. Buckingham, "David Hartley's Neural Vibrations and Psychological Associations," in *Brain, Mind and Medicine: Essays in Eighteenth-Century Neuroscience*, ed. H. Whitaker et al. (Boston, 2007), 177-190; Harry Whitaker & Yves Turgeon, "Charles Bonnet's Neurophilosophy," *ibid.*, 191-200. Besides models of vibrations and strings, the movement of fluids in a canal system played since Descartes a crucial role for the soul-body-interface. Jean-Paul Marat refers in *De l'homme* (1775-76) to the "soul" as an agent that moves a "fluid" in the "nerves." Cf. Jean-Paul Marat, *De l'homme, ou Des principes et des loix de l'influence de l'âme sur le corps et du corps sur l'âme*, 3 vols. (Amsterdam, 1775-1776), vol. 2, 83: "C'est du mouvement communiqué à ce fluide, combiné dans ses différens degrés de force avec l'élasticité des fibres & les divers organes affectés, que dépendent les phénomenes de l'influence du moral sur le physique." The "circulation" of liquids plays for Marat the same vivifying role in the fibre œconomy of as for Bonnet and Diderot the vibration of fibres. Cf. ibid., vol. 3, 1: "Le corps [organique] est une vraie machine hydraulique, composée d'un nombre prodigieux de vaisseaux & de diverses liqueurs. Tant que ces liqueurs circulent dans ces vaisseaux, l'animal est vivant: il ne l'est plus, dès qu'elles cessent entiérement d'y circuler."

63) Bonnet, *Méditations sur l'origine des sensations et sur l'union de l'ame et du corps*, in Bonnet, *Œuvres*, vol. 8, 391. Cf. Albrecht von Haller, Letter to Bonnet of the 6 September 1754; in Otto Sonntag, ed., *The Correspondence between Albrecht von Haller and Charles Bonnet* (Berlin, 1983), 49. Newton combined theories of electricity and vibration for the transmission of stimuli in organic bodies. Cf. Wes Wallace, "The Vibrating Nerve Impulse in Newton, Willis and Gassendi: First Steps in a Mechanical Theory of Communication," *Brain and Cognition*, 51 (2003), 66-94. For theories on animal electricity in the second half of the eighteenth century, see Walter Bernardi, *I fluidi della vita. Alle origini della controversia sull'elettricità animale* (Florence, 1992). Antonio Clericuzio, "The Internal Laboratory. The Chemical Reinterpretation of Medical Spirits in England (1650-1680)," in *Alchemy and Chemistry in the 16th and 17th*

In a letter of 25 June 1771, Bonnet challenges Haller, the "famous experimenter," to search for a single explanatory scheme for the initiation of movements in nerves and muscles through the combination of irritability and electricity:

> How does thus the contact of some fluid or solid bodies, of a sharp spine or any kind of acid, excite in the *Touts organiques* a more or less strong and continuous *Irritability*? ... I would like to have *Electricity* applied to *Irritability*, & and that much attention is paid to the effects of this application. It would be enough to bring an *electric fluid* on the recently detached limbs (*membres*) of their Subject. These experiments could be varied in thousands & thousands of different manners.[64]

The organic "network" thus is an electric "networking machine" (*ouvrage à réseau*)[65] of organized units on various levels, a system of a "nearly infinite" number of intermediate "chain-links" that appear in certain subordinated levels of order and that are regulated by a force that executes a preexisting plan. This networking system depends on the "enchained" microscopic structures of fibres or "threads" (*fils*) that form meshes, tissues, membranes, and canals for the transport of liquids and the constant assimilation and excretion of outer and inner parts.

In *Considérations sur les corps vivants* (1762), Bonnet points to the problem of organic parts that have no relation to nerves or a regulating soul, but still possess a kind of sensibility:

> I divide an insecte in the middle according to its size. It is evident that the part with the head is the only part that still conserves the *Self* or the *Personality*. There thus is no *Self* in the other part, because we have admitted that the Soul resides in the head; and yet, this parts seems to *sense*: it is capable of various movements. I have seen the half of an earth worm and sections of water worms that creep as

*Centuries*, ed. P. Rattansi & A. Clericuzio (Dordrecht, 1994), 51-83 focuses on the development of chemical explanations of the actions of animal spirits.

64) Letter from Bonnet to Haller of the 25 June 1771; in Sonntag, *Correspondence*, 946. Bonnet mentioned this idea already in the very beginning of his correspondence with Haller in a letter of 16 August 1754: "If you suppose that the *Esprits Animaux* have a Nature that is similar to the one of the *electric Fluid*, then the effects of the *irritations* of fibres correspond to the rubbing of electric Tubes." (ibid., 44) See also Bonnet, *Essai analytique*, 20-21 (§ 31).

65) Ibid., 65 (§ 97).

> if they would be complete worms. Further on, they seem to conserve all the inclinations (*inclinations*) that are proper for their species.[66]

Bonnet's reference to sensible organic parts is closely related to Diderot's fibre œconomy of organized bodies.[67] Like Bonnet, Diderot refers to a networking machine that is made out of interactive fibres. For Diderot, the basic property of these fibres is their sensibility. Through the sensible interaction of fibres, organic bodies can subsist without a specific soul-body-interface.

## 3. Diderot's Sensible Fibre Network

While Diderot uses Haller's *Primae lineae* (1747) as a source for the physical order of solid parts of organic bodies and Barthez's *Nouveaux éléments de la science de l'homme* (1778)[68] for the notion of sympathy between organs, he refers to Bordeu for his notion of a sensible fibre.[69] In *Recherches anatomiques sur la position des glandes,*

---

66) Bonnet, *Considérations*, 274 (§ 285). Bonnet continues (ibid., 275): "Ceux de mes Lecteurs qui ont lu les beaux Mémoires de M. de Haller sur l'*irritabilité*, entrevoient déja ce qu'on peut dire, pour tâcher à resoudre la difficulté dont il s'agit ici. On sait que l'irritabilité est cette propriété de la fibre musculaire, en vertu de la laquelle elle se contracte d'elle-même, à l'attouchement de tout corps, soit solide, soit fluide. C'est par elle, que le cœur, détaché de la poitrine, continue quelque tems à battre. C'est par elle, que les intestins, séparés du bas-ventre, et partagés en plusieurs portions, comme nos Vers, continuent pendant un tems, à exercer leur mouvement *péristaltique*. C'est par elle enfin, que les membres de quantité d'Animaux continuent à se mouvoir après avoir été séparés de leur tronc. Dira-t-on que ces portions d'intestins, qu'on voit ramper sur une table comme des Vers, sont mises en mouvement par une Ame qui réside dans leurs membranes?"

67) Cf. Denis Diderot, *Eléments de physiologie*, par Jean Mayer (Paris, 1964), 56-7: "La vie reste dans les organes separés du corps, où l'ame est-elle encore? Que devient son unité, son indivisibilité? ... Pourquoi ne pas regarder la sensibilité, la vie, le mouvement comme autant de propriétés de la matiere: puisqu'on trouve ces qualités dans chaque portion, chaque particule de chair?" For the influence of Bonnet's fibre model on Diderot, see Mayer, *Diderot*, 241 and 320-1.

68) The second, enlarged edition appeared in 1806.

69) In the fifteenth volume of the *Encyclopédie* (1765), Henri Fouquet (who like Bordeu was a medical doctor at Montpellier) defines "sensibility" as a general organic property that comprises both "irritability" and "sentiment" (*Encyclopédie*, vol. 15, 50): "L'*irri-*

*et sur leur action* (1752), Bordeu (professor at the medical faculty in Montpellier) is mainly interested in the secretion and excretion process of glands. Like Glisson, he wants to prove that their nerve-independent self-activity is based on a specific property that regulates the dynamic exchange of fluids between the inside and the outside of a gland. Bordeu calls this property "sensibility" or "sensitivity" (*sensibilité*). Through this property, fibres, tissues, organs, and systems of organs execute sequences of actions according to stimulus-reaction-schemes.

In *De natura substantiae energetica* (1677), Glisson focused on the autonomy of suborganic units as bladders that possess a faculty of "natural perceptions" (*perceptiones naturales*) through which they turn stimuli into actions.[70] Through "natural perceptions," bladders

---

*tabilité* n'est autre chose que la *mobilité* ou *contractilité* dont il a été question au commencement de cet article, & que nous avons dit être une des deux actions comprises dans l'exercice de la *sensibilité*; c'est toujours l'expression du sentiment; mais une expression violente, attendu qu'elle est le produit de la *sensibilité* violemment irrité par des *stimulus*; aussi est-elle quelquefois désignée sous le nom même de *stimulus* chez les Phisiologistes, ou sous celui de *fibre motrice*, &c." In the article 'Spasme' of the *Encyclopédie*, Jean-Joseph Menuret refers to the human as a "machine" in which each part is sensible or irritable: "Qu'est-ce que l'homme? ou pour éviter toute équivoque, que la méchanceté & la mauvaise foi sont si promptes à faire valoir; qu'est-ce que la machine humaine? Elle paroît à la premiere vûe; un composé harmonique de différens ressorts qui mûs chacun en particulier, concourent tous au movement général; une propriété générale particulierement restreinte aux composés organiques, connue sous les noms d'*irritabilité* ou *sensibilité*, se répand dans tous les ressorts, les anime, les vivifie & excite leurs mouvemens; mais modifiée dans chaque organe, elle en diversifie à l'infini l'action & les mouvemens; par elle les différens ressorts se bandent les uns contre les autres, se résistent, se pressent, agissent & influent mutuellement les uns contre les autres; cette commixture réciproque entretient les mouvemens, *nulle action sans réaction*. De cet anatagonisme continuel d'actions, résulte la vie & la santé." (*Encyclopédie*, vol. 15 (1765), 435b; thanks to Charles Wolfe for this reference).

70) While Glisson still explained the secretion processes of the gall bladder in *Anatomia hepatis* (1654) through nervous stimuli, he outlined in 1677 the metaphysics of a principle of individuation that relied on a self-organizing substance and its non-conscious perceptive or "irritable" properties in organically organized matter. For a detailed analysis of Glisson's work, see Temkin, "The Classical Roots"; Guido Giglioni, *The Genesis of Francis Glisson's Philosophy of Life*, PhD thesis, John Hopkins University (Baltimore, 2002); Karin Hartbecke, *Metaphysik und Naturphilosophie im 17. Jahrhundert. Francis Glissons Substanztheorie in ihrem ideengeschichtlichen Kontext* (Tübingen, 2006).

receive and react to stimuli. Bordeu transforms Glisson's concept of irritable "fibres" and "natural perceptions" into models of the self-activity of local agents that sense the stimulus without having nerves. Like perceptive agents, sensitive agents react to stimuli in a way that is specific for their energetic disposition. For Bordeu, sensitive stimulus-reaction-schemes are basic operative modi of the functional interaction of organs. He thus argues against the Hallerian dualism of nervous sensibility and muscular irritability. The interaction of organs that regulate different local regimes manifests itself physically in the "oscillations" (*oscillations*) of tissues and in a perpetual "trembling" (*tremblement, secousse*) of the whole body. Through these oscillations, organs communicate their different energetic states and "excitements."[71] However, Bordeu is ambivalent about the ontological status of his local agents and the distribution of regulative properties within the whole animal economy. He tentatively argues that the correlation of the different activities of local agents seems to require a single "maintaining force" (*force conservatrice*) that is not controlled by another agent. [72]

For Diderot, all regulative properties of organic bodies are related to the sensibility of fibres that results itself from the specific disposition of their "organized matter." Together with other organic parts as tissues and organs, fibres form a whole—the "entire animal" (*animal entier*). The organic parts of each organizational level have their own "live" (*vie*), emerging from the organization of the elements of the subordinated level, but they also interact with all the other parts. Further on, between the different parts and organized levels, there is not only a "sympathy" (*sympathie*) of different operational states, but also a hierarchy of regulating or controlling potentials—as between the brain and the other organic parts—, although subordinated levels, as the "system of organs," can subsist without a ruling centre.[73]

---

71) Cf. Théophile de Bordeu, *Œuvres complètes de Bordeu, précédées d'une notice sur sa vie et sur ses ouvrages*, par M. le Chevalier Richerand, 2 vols. (Paris, 1818), vol. 2, 802 and 830-831.

72) Cf. ibid., vol. 1, 163.

73) Cf. Diderot, *Eléments*, 286-287: "Voila donc des organes sensibles et vivans, accouplés, simpatisans et concourans à un même but sans la participation de l'animal

For his model of a sensible fibre network, Diderot combines aspects of Bordeu's secretion theory with Buffon's organic "molecules," La Mettrie's oscillating fibre machines and Maupertuis's theory of "molecular" epigenesis.[74] In *Le rêve de D'Alembert*, written in 1769 and posthumously published in 1830, Diderot introduces the organic body of a human as a "network" (*réseau*) of fibres:

> In a human body, the head, the feet, all the members, all the viscera, all the organs, the nose, the eyes, the ears, the heart, the lungs, the intestines, the muscles, the

---

entier." Cf. ibid., 27-28; Diderot, *Entretiens*, 135-136; and Kaitaro, *Diderot's Holism*, 102. Kaitaro (ibid., 100-101) also highlights the influence of Robert Whytt's *An essay on the vital and other involuntary movements of animals* (1751) on Diderot's model of the proper "life" of subunits of organic bodies.

74) For Bordeu's model of organic order, see Elizabeth Haigh, "Vitalism, the Soul, and Sensibility: The Physiology of Théophile Bordeu," *Journal of the History of Medicine*, 31 (1976), 30-41; Thierry Lavabre-Bertrand, *La philosophie médicale de l'école de Montpellier au XIXe siècle*, Thèse de doctorat, École pratique des Hautes Études (Paris, 1992); Dominique Boury, *La philosophie médicale de Théophile de Bordeu (1722-1776)* (Paris, 2004); and Cheung, *Res vivens*, 179-195, and "Concepts of Energy, Regulation, and Equilibrium From Stahl to Montpellier Vitalism," *Science in Context*, Special Issue: *Vitalism without Metaphysics? A Reevaluation of Medical Vitalism in the Enlightenment*, ed. Charles Wolfe, 2009, in print. Bordeu wrote the article 'Crisis' in the *Encyclopédie*. For the influence of Buffon on Diderot, see Timo Kaitaro, *Diderot's Holism: Philosophical Anti-Reductionism and Its Medical Background* (Frankfurt am Main, 1997), 65-66. For the general context of Diderot's model of organic order, see Herbert Dieckmann, "Théophile Bordeu und Diderots Rêve de d'Alembert," *Romanische Forschungen*, 52 (1938), 55-122; Marx W. Wartofsky, "Diderot and the Development of Materialist Monism," *Diderot Studies*, 2 (1952), 279-329; Mayer, *Eléments*, Introduction, xxxiii-xliii and 213-268; Roselyne Rey, "La théorie de la sécrétion chez Bordeu, modèle de la physiologie et de la pathologie vitaliste," *Dix-huitième siècle*, 23 (1991), 45-58, and "Diderot et les sciences de la vie dans l'Encyclopédie," *Recherches sur Diderot et sur l'Encyclopédie*, 18-19 (1995), 47-53; Bernard Baertschi, *Les rapports de l'âme et du corps* (Paris, 1992), 101-133; Kaitaro, *Diderot's Holism*, 84-8; François Duchesneau, "Diderot et la physiologie de la sensibilité," *Dix-huitième siècle*, 31 (1999), 195-216; Ileana Mihaila, "L'hylozoïsme de Diderot," in *Être matérialiste à l'âge des lumières. Hommage offert à Roland Desné*, ed. Béatrice Fink & Gerhardt Stenger (Paris, 1999), 185-196; Paolo Quintili, *La pensée critique de Diderot. Matérialisme, science et poésie à l'âge de l'Encyclopédie 1742-1782* (Paris, 2001), 295-353; and Andrew H. Clark, *Diderot's Part* (Hampshire-Burlington, 2008), 67-74. Fauvergue focuses on Diderot's interpretation of Leibniz's monadology, see Claire Fauvergue, *Diderot. Lecteur et interprète de Leibniz* (Paris, 2006).

> ears, the nerves, the membranes are really nothing but the coarse developments of a network that forms itself, grows, and produces a multitude of imperceptible threads.[75]

This fibre body is a "purely sensible system" (*système purement sensible*),[76] although some parts of the body, as hairs, are not sensible, and muscular fibres become "irritable."[77] The sensibility and irritability of a fibre depends only on its material organization - and not on the soul-body-interface (as in Haller's fibre body). In general, Diderot thinks that the fibre is itself a living "animal" or a "worm." As the main building block of the adult organic body, it is the constructive "principle of its entire machine." [78]

The organized structure of sensitive and irritable fibres, that is itself the result of the "assimilation" of embodied particles, transforms outer stimuli into patterns of vibrations which communicate—in a similar way as in Bordeu's model—irritations and movements throughout the entire body.[79] The fibre network, that forms the solid basis

---

75) Diderot, *Entretien*, 97.

76) Ibid., 103.

77) Diderot, *Eléments*, 65: "La fibre est sensible, les cheveux ne le sont pas: elle est irritable, quand elle est devenue musculeuse."

78) Cf. ibid., 65-66: "J'estime que la fibre est plus vraisemblablement de la chair ajoutée à de la chair, formant un tout continu, à peu près homogene, vivant. Je la regarde comme un animal, un ver. C'est cet être que l'animal qu'elle compose, nourrit. C'est le principe de toute la machine. Sa formation n'est que la formation d'un ver."

79) Cf. Diderot, *Entretien*, 39: " ... en mangeant, que faites-vous? Vous levez les obstacles qui s'opposaient à la sensibilité active de l'aliment. Vous l'assimilez avec vous-même; vous en faites de la chair; vous l'animalisez; vous le rendez sensible ..." Different grades of sensibility thus are "a product of organization" (ibid, 53). Cf. Diderot, Letter to Duclos, 10 October 1765, in Denis Diderot, *Correspondance*, ed. Georges Roth (Paris, 1959), vol. 5, 141: "L'animal est le laboratoire où la sensibilité, d'inerte qu'elle étoit, devient active." However, in *Refutation de l'ouvrage d'Helvétius intitulé 'L'Homme'* (written in 1773-74, unpublished), Diderot points to the still unsolved problem of an adequate explanation of the transformation of seemingly inert matter into sensible organic matter (*Œuvres*, vol. 2, 301): "J'invite tous les physiciens et tous les chimistes à rechercher ce que c'est que la substance animale, sensible et vivante. Je vois clairement dans le développement de l'œuf et quelques autres opérations de la nature, la matière inerte en apparence, mais organisée, passer par des agents purement physiques, de l'etat d'inertie à l'etat de sensibilité et de vie, mais la liaison nécessaire de ce passage m'echappe."

of organs of the adult body, is also the plastic material of its development. Fibre bundles "metamorphose" into various "tissues" (that interconnect organic parts) and "membranes" (that form "organs").[80] Like Schreiber and Bonnet, Diderot explains the "metamorphosis" through a point-line-tissue-scheme in the fictitious dialogue between Bordeu and Mademoiselle de l'Espinasse in *Le rêve de D'Alembert*:

> At the beginning, you were an imperceptible point, formed of the smallest molecules dispersed in the blood and lymph of your father and mother. This point became a slender thread, then a bundle of threads ... Each branch of this bundle of threads transformed merely by nutrition and by its structure (*conformation*) into a particular organ ...[81]

In Diderot's *Éléments de physiologie*, the material "organization" of fibres and their aggregates is made out of solid particles or "molecules" (*molecules*).[82] According to their proper "energies" or "forces," the molecules are composed of calcareous earth and a fluid element made of "water," "salt," "air," and "oil."[83] Diderot calls the fluid element "gluten" (*gluten*). Through the adhesive gluten, "molecules"

---

80) Cf. Diderot, *Eléments*, 68-70: " Le tissu est composé de fibres et de lames plus larges que longues qui ... unissent toutes les parties du corps humain, et font la fonction d'un lien qui les consolide sans les gener. ... La membrane fait les organes, comme la fibre fait le tissu cellulaire ..."

81) Diderot, *Entretien*, 103. Bordeu is speaking.

82) The text, probably written between 1778 and 1784, was first published in 1875. Cf. Mayer, *Eléments*, Introduction, xiii-xiv. All fibres are themselves composed of "simple fibres" (*fibres simples*) or "fibrillae" (*fibrilles*). Cf. ibid., 64. For Diderot's notion of the fibre œconomy of organic order in the *Pensées sur l'interprétation de la nature* (1753), see Varloot, in Denis Diderot, *Œuvres complètes*, ed. H. Dieckmann et al., 33 vols. (Paris, 1975-), vol. IX (1981), Introduction, 3-14; Geoffrey Bremner, "Les *Eléments de physiology* et le sens de la vie," in *Diderot. Les dernières années 1770-84*, ed. P. France & A. Strugnell (Edinburgh, 1985), 81-91; Roselyne Rey, "Dynamique des formes et interprétation de la nature," *Recherches sur Diderot et sur l'Encyclopédie*, 11 (1991), 49-62; and Duchesneau, "Diderot et la physiologie."

83) Cf. Diderot, *Œuvres*, vol. 17 (1987), 17 (*Principes philosophiques sur la matière et le mouvement*): "Toute molécule doit être considérée comme actuellement animée de trois sortes d'actions, l'action de pesanteur ou de gravitation, l'action de sa force intime et propre à sa nature d'eau, de feu, d'air, de soufre; et l'action de toutes les autres molécules sur elle; et il peut arriver que ces trois actions soient convergentes ou divergentes."

also interconnect and form fibres. The different properties of the "molecules" thus "assimilate" into the specific sensibility of various fibre types.

Haller already explained the elastic and irritable properties of muscles (but not the sensibility of organic bodies) through an element called gluten.[84] For Diderot, gluten acts, as the entire fibre, according to a reflex mechanism of stimuli-dependent contractions and dilatations or "actions" and "reactions." Combined with the inner dynamics or "restlessness" (*inquiétude*) of the "molecules," this

---

[84] Cf. Diderot, *Eléments*, 9-10 and 63-5. Diderot (ibid., 10) refers to Pierre Jean Macquer's and Guillaume Rouelle's chemical experiments on the components of flour: " ... M. Rouelle à Paris, M. Macquer et les plus habiles de nos chimistes ... trouverent que l'amidon ne contenait, pour bien dire, que les parties vegetales de la farine; qu'en enlevant l'amidon, il restait un *gluten* qu'ils appellerent *vegeto-animal*: toutes ses parties sont si rapprochées, si liées entre-elles, qu'on ne peut les separer: Quand on le tire, il s'étend dans tous les sens; et quand on l'abandonne, il se replie sur lui même, et reprend sa premiere forme, comme fait le tissu de la peau qui tour-à-tour se reserre."; For Haller, see Robert E. Schofield, *Mechanism and Materialism. British Natural Philosophy in An Age of Reason* (Princeton, 1970), 202; Baertschi, *Les rapports de l'âme*, 37-8; Guido Giglioni, "What Ever Happened to Francis Glisson? Albrecht Haller and the Fate of Eighteenth-Century Irritability," *Science in Context* 21 (2008), 465-93, 476; and Albrecht von Haller, *A Dissertation on the Sensible and Irritable Parts of Animals* (London, 1755, reprint Baltimore, 1936), 40-42: "But the muscular fibres being composed of earthy particles and a glutinous mucus, it may be asked in which of these Irritability resides. It appears most probably to recide in the latter, because this when it is pulled endeavours to shorten itself; whereas on the contrary, dry earth never changes its figure of itself, and being extreamly brittle, when its parts are separated, they constantly remain so ... What therefore should hinder us from granting Irritability to be a property of the animal *gluten,* the same as we acknowledge attraction and gravity to be properties of matter in general, without being able to determine the cause of them. Experiments have taught us the existence of this property, and doubtless it is owing to a physical cause which depends upon the arrangement of the ultimate particles, though the experiments that we can make are too gross to investigate them." For the debate between Haller and Whytt on the role of an irritable glue that causes the contraction of muscles, see Eugenio Frixione, "Irritable Glue: The Haller-Whytt Controversy on the Mechanism of Muscle Contraction," in *Brain, Mind and Medicine*, ed. Whitaker et al., 115-124. Schofield (*Mechanism and Materialism*, 148) points also to Boerhaave's notion of gluten as an adhesive organic material. For the problem of the "elasticity" of the fibre, see Hisao Ishizuka, "The Elasticity of the Animal Fibre: Movement and Life in Enlightenment Medicine," *History of Science*, 44 (2006), 435-468.

movement results into vibrations that are communicated throughout the body.[85] From the vibrating fibre œconomy emerge the three organizational levels that characterize "life" in animals and humans: "The life of the entire animal, the life of each of its organs, the life of the molecule."[86]

The organization of organic bodies is for Diderot emergent, and yet it results with necessity from the vibrating movements of action and reaction whose agents are "molecules" and their organic structures, the "fibres." The animal body is a fibre "system of actions and reactions" (*un sisteme d'actions et de reactions*) that "causes the forms of inner organs and membranes." Diderot thus transforms Bordeu's notion of the proper "lives" of organs and their systemic interaction into a process of the formation of organized organic bodies. Life is nothing else than "a sequence of actions and reactions."[87]

---

85) Cf. Diderot, *Eléments*, 8; and *Pensées sur l'interprétation de la nature*, in *Œuvres*, vol. 9 (1981), 84. For Diderot's notion of "living molecules" (*molécules vivantes*), its relation to Leibniz's monads and Maupertuis' minima, see Wartofsky, *Diderot*, 313; Jean Deprun, *La Philosophie de l'inquiétude en France au XVIIIème siècle* (Paris 1979), 202-205; H.H. Kubbinga, *L'histoire du concept de «molécule»*, 3 vols. (Paris, 2002), vol. 1, 385-437; Jean-Claude Bourdin, *Diderot. Le Matérialisme* (Paris, 1998); and Charles Wolfe's contribution in this volume.

86) Diderot, *Eléments*, 27. Cf. ibid., 23: "Il y a l'infinie varieté des chocs relatifs à la masse, l'infini variété des chocs relatifs à la vitesse, l'infinie variété d'une qualité phisique, l'infinie variété des effets combinés d'une seconde, d'une troisième, d'une multitude de qualités phisiques; et tous ces infinis se combinent ecnore avec la variété des organes, et peut-être des parties de l'animal." In *Le rêve de D'Alembert*, Diderot refers also to the analogy of a "swarm of bees" to characterize the specific unity of interactive organic parts. Maupertuis already used this metaphor in his *Dissertation d'Erlangen* (1751) and Bordeu in *Recherches anatomiques sur les différentes positions des glandes et leur action* (1752). Cf. Dieckmann, "Théophile Bordeu," 86-87; Vernière's commentary in Denis Diderot, *Œuvres philosophiques*, textes établis, avec introductions, bibliographies et notes, par Paul Vernière (Paris, 1956), 291-2, footnote 1; Diderot, *Entretien*, 71-72; Kaitaro, *Diderot's Holism*, 84-94; Charles Wolfe, "Machine et organisme chez Diderot," *Recherches sur Diderot et sur l'Encyclopédie* 26 (1999), 223-224; and Thomas Becker, *Mann und Weib—schwarz und weiß. Die wissenschaftliche Konstruktion von Geschlecht und Rasse 1600-1950* (Frankfurt am Main, 2005), 206-207.

87) Cf. Diderot, *Entretien*, 70-71 and 94: "La vie, une suite d'actions et de réactions ... Vivant, j'agis et réagis en masse ... mort, j'agis et réagis en molécules." (Mademoiselle de l'Espinasse is speaking about Diderot's dream); and *Eléments*, 61: "Le corps animal est un sisteme d'actions, et de reactions: actions et réactions sont les causes des

If the fibre œconomy of organized matter is combined with a "soul" in human bodies, the thinking soul acts as a "spider" that senses the vibrations of its net.[88] However, for Diderot, the soul-like "spider" seems not an entity that is categorically different from the rest of the organized body. Rather, Diderot suggests that the reflexive and mnemonic faculties of the soul, as well as the "instincts," are an emerging property of the organic fibre œconomy itself (although the different levels of organization are not directly reducible to combinations of simple dispositions of matter). [89] The spider is part of

---

formes des visceres, des membranes." Cf. Bordeu, *Œuvres*, vol. 1, 187: " ... les organes du corps sont liés les uns avec les autres; ils ont chacun leur district et leur action. Les rapports de ces actions, l'harmonie qui en résulte font la santé."; and Vernière's commentary in Diderot, *Œuvres philosophiques*, 292, footnote 1: "C'est donc Bordeu qui a fourni à Diderot cette opposition de la vie générale et de la vie paerticulière des organes qu'il reprendra plus loin. Mais il ne doit qu'à lui-même cette extension de l'image au problème des origines de l'être vivant." For Diderot's various systems of action and reaction, see Jean Starobinski, *Action et réaction. Vie et aventures d'un couple* (Paris, 1999), 53-144.

88) Cf. Diderot, *Entretien*, 97: "Les fils sont partout; il n'y a pas un point à la surface de votre corps auquel ils n'aboutissent; et l'araignée est nichée dans une partie de votre tête que je vous ai nomé, les méninges, à laquelle on ne saurait presque toucher sans frapper de torpeur toute la machine." For the difference between the "sensations" (*sensations*) of a soul and the "sensibility" (*sensibilité*) of fibres, see ibid., 60. Montesquieu already referred to the spider-metaphor in his *Essai sur les causes qui peuvent affecter les esprits et les caractères*. According to Roger Caillois, the text has been written before the *Esprits des lois* (1748), but it was not published before 1892. Cf. Charles-Louis Montesquieu, *Œuvres complètes*, texte présenté et annoté par Roger Caillois, 2 vols. (Paris, 1951), vol. 2, 49: "L'âme est, dans notre corps, comme une araignée dans sa toile. Celle-ci ne peut se remuer sans ébranler quelqu'un des fils qui sont étendus au loin, et, de même, on ne peut remuer un de ces fils sans la mouvoir. On ne peut toucher un de ces fils qu'il en remue quelque autre, qui lui répond. Plus ces fils sont tendus, mieux l'araignée est avertie. S'il y en a quelques-uns de lâches, la communication sera moindre de ce fil à l'araignée ou de ce fil à un autre fil, et la providence de l'araignée sera presque suspendue dans sa toile même." See also Catherine Glyn Davies, *Conscience as consciousness: the idea of self-awareness in French philosophical writing from Descartes to Diderot* (Oxford, 1989), 126.

89) Cf. Diderot, *Eléments*, 32: "L'instinct animal est un enchaînement necessaire de mouvemens consequens à l'organisation et aux circonstances, par l'effet des quelles l'animal execute sans nulle deliberation, independamment de toute experience, une longue suite d'operations conformes à sa conservation; si cela ne se pouvait, l'animal ne serait pas."; and ibid., 59: „L'animal est *un*, et c'est peut être cette unité qui

a loom-like apparatus that spins the material of which organic bodies are made and through which they live.[90]

The life sustaining activities of organs do not necessarily depend on the influence of a regulating centre. Fibres are contractible in dead animals, and their contractibility always represents a kind of sensibility, that is to say a faculty to receive impressions of other entities in order to assimilate them and to react.[91] This faculty is a property of organized matter or "flesh" (*chair*).[92] It distinguishes

---

constitue l'ame, le soi, la conscience à l'aide de la memoire." For Diderot, animals are more influenced by the senses than by reason. Cf. Diderot, *Réfutation suivie de l'ouvrage d'Helvétius intitulé* L'Homme, *Œuvres*, vol. 24, 557: "Pourquoi l'homme est-il perfectible et pourquoi l'animal ne l'est pas? L'animal ne l'est pas, parce que sa raison, s'il en a une, est dominée par un sens despote qui la subjuge. Toute l'âme du chien est au bout de son nez; et il va toujours flairant. Toute l'âme de l'aigle est dans son œil, et l'aigle va toujours regardant ... Mais il n'en est pas ainsi chez l'homme. Il est entre ses sens une telle harmonie, qu'aucun ne prédomine assez sur les autres pour donner la loi –a son entendement; c'est son entendement au contraire, ou l'organe de sa raison qui est le plus fort." For the problem of the emergence of reflexive properties and the animal-human-distinction, see also Kaitaro, *Diderot's Holism*, 76-79; Bourdin, *Diderot*, 36-43; Quintili, *La pensée critique de Diderot*, 337-340 and 352; Fauvergue, *Diderot*, 186-188; and Ann Thomson, *Bodies of Thought. Science, Religion, and the Soul in the Early Enlightenment* (Oxford-New York, 2008), 219-222.

90) Cf. Diderot, *Eléments*, 22-3: " ... Que serait-ce qu'un metier de la manufacture de Lion si l'ouvrier et la tireuse faisant un tout sensible avec la trame, la chaine, et le simple? Ce serait un animal semblable à l'araignée qui pense, qui veut, qui se nourrit, se réproduit et ourdit sa toile." In the *Encyclopédie* (vol. 1 (1751), XLV), a short note indicates that two workers showed Diderot how a loom produces cloth, especially "velour": "MM. Bonnet & Laurent, ouvriers en *Soie*, ont monté & fait travailler sous les yeux de M. Diderot, un métier *à velours*, &c. & un autre en *étoffe brochée*: on en verra le détail à l'article VELOURS." Cf. Mayer, *Eléments*, 22-23, footnote 3. Shortly after the lemma "Velours" (vol. 16 (1965), 882-907), Diderot compares in the lemma "Veloutée" the "tunic of the intestines" (*tunique des intestines*) with the structure of the textile "velour" (vol. 16, 908): "On la nomme *tunique veloutée* ou *tunique villeuse*, à cause de certains poils semblables à ceux du velours, dont elle paroît couverte ... Elle sert à couvrir les orifices des vaisseaux, à les défendre contre les effets nuisibles des matieres qui peuvent passer ou être contenues dans le conduit intestinal, & à transmettre ses impressions à la tunique nerveuse." For a detailed discussion of Diderot's spider metaphor, see Ilie, *The Age of Minerva*, vol. 2, 142-49.

91) Cf. Diderot, *Eléments*, 28 and 65.

92) Cf. ibid., 66: "Tous les elemens que l'on trouve par l'analise dans la fibre, forment la chair; cette espece de chair ainsi ordonnée forme la fibre, et la fibre est organisée fibre

living and non-living beings.[93] Only living beings are made of and exist through sensible flesh and its fibre œconomies, and yet non-organic matter can be transformed into flesh.

## 4. Concluding Remarks

In a long-term transformation that begins in Antiquity but takes a crucial turn in the Renaissance anatomies of canal systems and muscle structures, the "fibre" becomes the basic building-block and self-active unit of organic bodies in which physiological processes take place on various structural levels of interwoven forms as tubes and organs.

In Bonnet's and Diderot's fibre œconomies, the self-active, regulating properties of fibre-agents and their material structures, that reach from tissues and membranes to apparati of organs, are united within the concrete whole of individual organized "systems" or "networks." However, the explanatory scheme of fibre systems is very different in Bonnet's and Diderot's models of organic order. For Bonnet, the fibre is not only the first operative element that constitutes the reproductive germ, but also the highly differentiated product of the germ within organic œconomies that characterize plants, animals and humans. Bonnet focuses on a process of weaving that begins with germination and ends in the fibre network of the adult body. Within each reproductive fibre-germ-unit, he explains organic development through a dynamic plan-structure-interface that is centered upon a weaving agent. Organic bodies thus exist as living

---

en consequence, comme l'arbre de Diane est arbre de Diane." In the *Encyclopédie*, the process that produces the Tree of Diane is defined as an alchemical fusion of "silver," "mercury," and the "spirit of nitrate" (*Encyclopédie*, ed. by Diderot and d'Alembert, vol. 1 (1751), 590): "Végétation métallique artificielle, dans laquelle on voit un *arbre* se former & croître peu-à-peu du fond d'une bouteille pleine d'eau. Cette opération se fait par le mêlange de l'argent, du mercure & de l'esprit de nitre qui se crystallisent ensemble en forme d'un petit arbre." Cf. Diderot, *Eléments*, 66, note 1. The Tree of Diane is also the name for a still existing plant that has been planted by Diane de Poitiers in the park of Clayes-sous-Bois in 1556.

93) Cf. Diderot, *Eléments*, 71: "Tout concourt donc à produire une sorte d'unité qui n'existe que dans l'animal … "

looms with regulating properties and a central weaving agent. Their basic active unit is the fibre in the fibre-germ-unit. Unlike Bonnet, Diderot argues that the regulating properties of fibres and their networks merely result from the specific "energies" or "forces" of interactive heterogeneous "molecules" that are composed of earthy matter and gluten. For him, the unity of the "system of organs" and of the "entire animal" represent emerging organizational levels that are unique to living beings and that cannot be reduced to simple cause-and-effect-chains of parts.

The human soul is for Bonnet not an integrative part of the fibre networks, while Diderot thinks of it as an emerging entity that corresponds as a kind of consciousness to the soul-like function of brains in animals.[94] However, both refer to mechanical analogies of vibration to explain its operational mode. Only in his later writings does Bonnet tend to an electrical theory of the transmission of stimuli. While Bonnet uses the analogy of a soul that plays like an organist on the vibrating fibre œconomy of living bodies, Diderot refers to the organic body as a clavecin in which fibre complexes are at the the same time passive mechanisms that receive vibrations and agents that sense, memorize and produce them.[95] Diderot also compares the soul (as an integrative part of the living body) to a spider that sits in the brain of an organic web.

After 1800, the fibre gradually becomes a transitional organizational level of the organic order between "molecules" or "globules" and "tissues." While the tissue replaces the fibre as the main operational element in the catalogue of twenty-one tissue types of Bichat's *Anatomie générale* (1801), Henry Dutrochet's and Jean-Louis Prévost's search for "globules," that compose and form the microstructure of fibres, intensifies discussions about the first living elements

---

94) For Bonnet, the human soul is spontaneous and free in its actions. For Diderot, the soul is determined by the material body within which it acts. Cf. Bonnet, *Essai de psychologie*, 120-1; and Diderot, *Eléments*, 59-60.

95) Cf. Diderot, *Entretiens*, 50: "Nous sommes des instruments doués de sensibilité et de mémoire. Nos sens sont autant de touches qui sont pincées par la nature qui nous environne, et qui se pincent souvent elles-mêmes; et voici, à mon jugement, tout ce qui se passe dans un clavecin organisé comme vous et moi."

of organic bodies.[96] In his *Handbuch der Physiologie* (1833), Johannes Müller points to the problem of deciding if the seemingly infinitely small "primitive fibres" (*Primitivfasern*) of organic bodies are themselves composed of "organic molecules."[97] In the same decade, Matthias Jacob Schleiden and Theodor Schwann make substantial contributions to the experimental and theoretical foundations of Rudolf Virchow's new unifying paradigm of the growing life sciences, the paradigm of *omnis cellula e cellula*. This paradigm, first formulated by Virchow in 1855, has its own history of explanatory schemes and analogies.[98] But the discourses of the fibre and the cell merge in the 1830s. In 1838, Schleiden defines the "double life" of a plant cell: one for its own "development" and another for its interactions as an "integrating part" of the organism.[99] One year later, Schwann compares the double life of the plant cell with that of an animal cell.[100] For him, all living beings are formed by and composed of cells. For a similar purpose, but within a different discursive and historical setting, Bonnet compared in 1762 in the first chapters of *Considérations sur les corps organisés* the development of plants and animals as fibre-germ-units.

---

96) Cf. Walter Wilson, "Dutrochet and the Cell Theory," *Isis*, 37 (1947), 14-21; and John V. Pickstone, "Globules and Coagula: Concepts of Tissue Formation in the Early Nineteenth Century," *Journal of the History of Medicine*, 28 (1973), 336-356.

97) Cf. Johannes Müller, *Handbuch der Physiologie des Menschen für Vorlesungen*, 2 vols. ([1st ed. 1833], 3rd ed. Koblenz, 1837), vol. 1, 9.

98) Cf. Stephen L. Jacyna, "The Romantic Programme and the Reception of Cell Theory in Britain," *Journal of the History of Biology*, 17 (1984), 13-48; Duchesneau, *Genèse*; and Ohad Parnes, "The Envisioning of Cells," *Science in Context*, 13 (2000), 71-92.

99) Cf. Matthias J. Schleiden, "Beiträge zur Phytogenesis," *Archiv für Anatomie, Physiologie und wissenschaftliche Medien*, Jahrgang 1838, 137-176, 138.

100) Cf. Theodor Schwann, *Mikroskopische Untersuchungen über die Übereinstimmung in der Struktur und dem Wachsthum der Thiere und Pflanzen* (Berlin, 1839).

# Transhumane Physiologie.
# Bilder und Praktiken des Reflexes
# (Thomas Willis, Robert Whytt, Marshall Hall)

**Yvonne Wübben**
*Freie Universität Berlin**

**Abstract**
The essay examines the function of visualizations and practices in the formation of the reflex concept from Thomas Willis to Marshall Hall. It focuses on the specific form of reflex knowledge that images and practices can contain. In addition, the essay argues that it is through visual representations and experimental practices that technical knowledge is transferred to the field of human reflex physiology. When using technical metaphors in human physiology authors often seem to feel obliged to draw distinctions between humans, machines and animals. On closer scrutiny, these distinctions sometimes fail to establish firm borders between the human and the technical.

**Keywords**
reflex physiology, experimental practices, technical knowledge, scientific images

Abbildung 1 zeigt einen modernen Reflexhammer, der 1910 in den deutschsprachigen Raum eingeführt und nach seinem Konstrukteur, dem Hamburger Neurologen Ernest L. Trömner (1868-1930), benannt wurde.[1] Wie das Stethoskop und andere mobile Diagnose-Instrumente kann der Hammer als Emblem der Professionalisierung verstanden werden. Er steht für eine diagnostische Praxis, für Reflexprüfungen,

* Institut für Deutsche und Niederländische Philologie, Freie Universität Berlin, Habelschwerdter Allee 45, 14195 Berlin, Germany (ywuebben@zedat.fu-berlin.de).

[1] E. Troemner, „Ein neuer Reflexhammer, zugleich ein Beitrag zur Prüfung von Sehnenreflexen" in *Berliner Klinische Wochenschrift* 21 (1910), 997-998 sowie ders.: „Ueber Sehnen- resp. Muskelreflexe und die Merkmale ihrer Schwächung und Steigerung", in *Berliner Klinische Wochenschrift*, 50 (1913), 1712-1715.

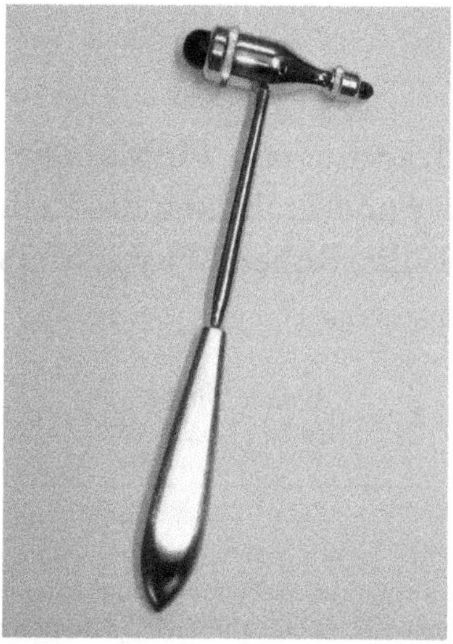

Abbildung 1: Reflex-Hammer, Modell Trömner.

die seit dem Ende des 19. Jahrhunderts zum festen Bestand neurologischer Statuserhebungen gehören. In seiner 1880 publizierten Schrift *The Diagnosis of Disease of the Spinal Cord* illustrierte der englische Neurologe Wilhelm Gowers (1845-1915) eine solche Reflexprüfung und bildete damit für die klinische Diagnostik wichtiges know-how ab (siehe Abbildung 2).[2]

Die eigentliche Pointe besteht jedoch nicht in der medizinisch-diagnostischen Praktik. Sie besteht vielmehr im Umstand, dass jede Prüfung nicht nur einen Reflex auslöst, sondern auch Wissen über den Reflex aktualisiert. Das Wissen vom Reflex ist seinerseits historisch gewachsen, so dass man zunächst ebenso pointiert wie provokant fragen möchte: War zuerst die Praktik oder zuerst das Wissen über den Reflex da? Blickt man auf die Geschichte des Reflexes, erweist sich die Praktik als eine wichtige Dimension des Begriffs, die auch mit seiner anthropotechnischen Geschichte zu tun hat. Reflex-Praktiken etablieren Gemeinsamkeiten zwischen Mensch und Tier, insofern der Reflex zuerst beim Tier und dann auch beim Menschen

[2] William R. Gowers, *The Diagnosis of Disease of the Spinal Cord* (London, 1880).

Abbildung 2: Gowers, Diagnosis, 1880.

ausgelöst wurde und in der Regel nicht von höheren neuronalen Strukturen abhängig ist. Die seit Marshall Hall (1790-1857) gültige Bestimmung des Reflexes lautet:

> This [new] property is characterized by being excited in its action, and reflex in its course: in every instance in which it is exerted, an impression made upon the extremities of certain nerves is conveyed to the medulla oblongata or the medulla spinalis, and is reflected along other nerves.[3]

Hall begreift den Reflex als ein dezentrales und exzentrisches Ereignis, das vor allem von der medulla oblongata und der medulla spinalis vermittelt wird. Für diese Begriffsspezifizierung waren, so die These, zunächst Praktiken und Bilder zentral. Denn ebenso wie die klinische Prüfung Reflexwissen im 19. Jahrhundert verstetigt hat, haben Praktiken und Bilder zur Herausbildung von Marshall Halls Reflexbegriff beigetragen: und zwar nicht nur deshalb, weil sie Wissen illustrieren und der Evidenz-Erzeugung dienen,[4] sondern auch, weil

---

3) Marshal Hall, „On the Reflex Function in the Medulla Oblongata and Medulla Spinalis", in *Philosophical Transactions of the Royal Society*, 123 (1833), 635–665, 652.

4) Zu dieser Funktion siehe: Peter Machamer, „17th Century Demonstrations", in *Wissensideale und Wissenskulturen in der frühen Neuzeit*, ed. Wolfgang Detel und Claus Zittel (Berlin, 2002), 23-37.

ihnen als Medien oder Handlungen eigenes Wissen zukommt.[5] Ohne die experimentellen Praktiken und frühneuzeitlichen Bilder wäre, so möchte ich behaupten, der Reflex weder begrifflich geworden noch in die klinische Diagnostik eingegangen.

Diverse für den Reflex zentrale Bilder stammen aus dem Bereich der Technikgeschichte, etwa der Hydraulik und der Pyrotechnik der Frühen Neuzeit. Die für die Herausbildung des Reflexbegriffs entscheidenden Praktiken vermittelten sich wiederum im Tierversuch. Es hängt auch mit diesen Kontexten, d.h. mit der medialen Geschichte des Reflexes zusammen, dass er auf Mensch, Tier und Maschine gleichermaßen anwendbar ist. Eine auf Bilder und Praktiken bezogene Reflexgeschichte zeigt, wie über diese Medien Gemeinsamkeiten zwischen den keinesfalls als ahistorisch zu verstehenden Entitäten Mensch, Tier und Maschine gestiftet werden.

Für die frühneuzeitlichen Kontexte lassen sich diese Zusammenhänge in drei Schritten skizzieren: Der erste Teil des Aufsatzes wird sich mit Thomas Willis (1621-1675) und dessen Traktat über die Muskelbewegung befassen, der eine zentrale Komponente des späteren Reflex-Begriffs formuliert. Diese Komponente erschließt sich erst im Blick auf die pyrotechnischen Bilder der Zeit. Der zweite Teil untersucht die Reizexperimente Robert Whytts (1714-1766), der in seinem Traktat über die unwillentliche Muskelbewegung Praktiken erwähnt, die den Reflex-Begriff ermöglichen, weil sie der Reflexhandlung, so meine These, eine neue zeitliche Dimension verliehen. Diese Praktiken werden meist im Tierversuch eingeübt und legen die Vergleichbarkeit von Tier und Mensch nahe. Am Tier wird somit Wissen erprobt, das später in den Reflex, d.h. auch in die klinische Reflexprüfung eingeht. Im Zentrum des letzten Teils steht der bereits zitierte Marshall Hall, dessen Versuche schließlich zur Ausbildung anthropotechnischer Organismen führen.

---

5) Bruno Latour, „Drawing Things Together. Die Macht der unveränderlich mobilen Elemente“ in *Anthology. Ein einführendes Handbuch zur Akteur-Netzwerk-Theorie*, ed. Andréa Belliger und David J. Krieger (Bielefeld, 2006), 259-307, hier 260.

## 1. Instantanität: Thomas Willis' Motus Reflexus

In seiner Schrift *De motu musculari* (1680) hat der Oxforder Arzt Thomas Willis den Terminus „motus reflexus" eingeführt und ist deshalb vom Reflexhistoriker Georges Canguilhem 1955 zum Begründer des Reflexes ernannt worden.[6] Bei Willis heißt es: „motus utriusque generis est reflexus, qui silicet a sensione praevia tanquam causa evidenti aut occasione immediatius dependens, illico retorquetur."[7]

Wills nennt eine Bewegung dann reflektiert, wenn ihr eine Empfindung vorausgeht und sie zum Ort der Empfindung „zurückgeworfen" (retorquere) wird. Er benutzt den Begriff aber anders als Marshall Hall nicht strukturfunktional, d.h. er bezieht ihn nicht auf konkrete anatomische Verhältnisse, etwa die medulla spinalis. Ähnlich wie zuvor Galen und Aristoteles nimmt Willis vielmehr eine abstrakte Bewegungsklassifikation vor.[8] Er unterscheidet vier Typen von Bewegungen nach folgenden Kriterien: erstens nach dem Ursprung der Bewegung, zweitens danach, ob sie spontan, d.h. willentlich oder unwillentlich erfolgen, drittens, ob sie direkt oder rückgebogen sind, und viertens danach, ob sie von nervösen oder anderen Strukturen ausgehen.[9] Reflexbewegungen werden dadurch gekennzeichnet, dass ihnen eine „Ursache" (causa) bzw. eine „Gelegenheit" (occasio) vorausgeht und dass sie aus zwei an sich distinkten Ereignissen zusammengesetzt sind. Als Beispiel für eine solche Reflexbewegung gibt Willis etwa das Kratzen auf einen Juckreiz an:

> As to the original beginning of Motion, we shall take notice, that that which proceeds from the Brain, with a knowing and auspicious appetite, may be called Spontaneous or Voluntary; but that which is wont to be excited from the Cerebel where

---

6) Georges Canguilhem, *Die Herausbildung des Reflexbegriffes im 17. und 18. Jahrhundert.* A. d. Frz. übers. u. durch e. Vorw. eingel. v. Henning Schmidgen (München, 2008).

7) Thomas Willis, *De motu musculari* [1644], zit. n. Canguilhem, *Die Herausbildung des Reflexbegriffes*, 86: „Jede dieser beiden Bewegungen ist reflektiert, wenn sie unmittelbar von einer vorhergehenden Empfindung als offensichtlicher Ursache oder Gelegenheit abhängt und dorthin zurückgeworfen wird."

8) Aristoteles, *De motu animalium,* 703b2-21.

9) Thomas Willis, *Five Treatises, The Second Medical and Physical Discourse, Of Musculary Motion* (London, 1681), 34.

> the Law of Nature presides, such as are Respiration, the Pulse, with many others, may be called merely Natural, or Involuntary; either of these is either direct, which is stirred up to it self, or primarily from this or that beginning, as often as the appetite requires this or that thing [...] or the motion of either kind is reflected, to wit, which depending on a previous sense more immediately, as an evident cause or occasion, is presently retorted, so a gentle titillation of the Skin causes a rubbing of it, and the more intense heats of the Praecordia stir up the Pulse and Respiration.[10]

In diesem Beispiel entspricht der Juckreiz der Empfindung und das Kratzen der unmittelbar auf ihr folgenden Bewegung. Canguilhem hat die oben zitierte Passage als Vorwegnahme des Reflexbegriffes gedeutet und das Zurückwerfen als die von Marshall Hall beschriebene Impulsumschaltung am Rückenmark verstanden.[11] Im Text fehlen allerdings Anhaltspunkte, die eine solche letztlich anatomisch-physiologische Deutung rechtfertigen, da Willis das Zurückwerfen weder physiologisch noch strukturfunktional versteht. Die Bedeutung des ‚retorquere' wirft vielmehr die Frage danach auf, inwiefern die Bewegung eigentlich zurückgeworfen wird und wie man sich diesen Vorgang konkret, d.h. physiologisch vorzustellen hat. Blicken wir zur Klärung dieser Zusammenhänge nochmals auf die schon zitierte Schrift *De motu musculari*, der das Zitat entnommen ist. Ein weiteres von Willis zentralen Reflex-Beispielen ist die Veränderung des Herzschlages auf Wärme, also jenes Typs von Bewegung, den Aristoteles mit der Furcht in Zusammenhang gebracht hat. Den Vorgang der Kontraktion erklärt Willis sehr wohl physiologisch bzw. chemisch, indem er ihn mit einer Explosion vergleicht.[12]

Über den pyrotechnischen Vergleich stellt er einen unmittelbaren Bezug zwischen dem als *motus reflexus* bezeichneten Phänomen – der Herzbeschleunigung – und dem zugrunde liegenden physiologischen Vorgang her. Obschon auch die Reflexbewegung demnach einer Explosion gleicht, liefert dieser Vergleich über den Reflex dennoch keinen unmittelbaren Aufschluss, denn er gilt für jede Muskelbewegung gleichermaßen, d.h. auch für voluntative Bewegungen,

10) Ebd.

11) Canguilhem, *Die Herausbildung des Reflexbegriffes*, 86.

12) Willis, *Of Musculary Motion*, 45f.

die über das Gehirn gesteuert und durch Nervengeister vermittelt werden. Daher ist hier nochmals zu fragen, was der Ausdruck ‚zurückwerfen', der die Reflexbewegung spezifisch von anderen Bewegungen unterscheidet, eigentlich bedeutet.

Hinweise auf die Bedeutung von ‚retorquere' dürfte der Kontext des Vergleichs liefern. Wenn Willis die Muskelkontraktion mit der Explosion parallelisiert, dann greift er auf die Pyrotechnik seiner Zeit zurück, wie vor ihm bereits Francis Glisson. Es ist wahrscheinlich, dass Willis neben theoretischen Erörterungen auch zahlreiche bildliche Darstellungen der Pyrotechnik kannte. Die Bedeutung der Pyrotechnik für Willis' Reflexbegriff erschließt sich, so meine These, überhaupt erst durch die Einbeziehung der zeitgenössischen Bilder und ihrer diagrammatischen, d.h. umrisshaften und schematischen Aspekte. Die Bilder zeigen nämlich mehr als Explosionen. Sie stellen zudem die Ausbreitung der Explosion dar und ergänzen damit einen wesentlichen für den Reflex zentralen Aspekt, der sich über die Bedeutung von ‚retorquere' zunächst nicht unmittelbar erschließt. Abbildung 3 zeigt eine für Willis zeitnahe Visualisierung eines Feuerwerkes.

Die dargestellte Explosion ist der Nautik entnommenen. Abbildung 3 zeigt uns ein brennendes Schiff, von dem Feuersterne wie wachsende Pflanzen in die Luft ragen. In der Bildmitte befindet sich das Zentrum des Feuerwerkes, das mit den beiden, jeweils am Rand angeordneten Sonnen ikonographisch parallelisiert wird. Mit der Muskelkontraktion hat die Explosion zunächst die Unmittelbarkeit gemein. Wie jene erfolgt sie unmittelbar und breitet sich schnell aus. Im Bild erscheint aber nicht nur die instantane Bewegung der Explosion. Die dargestellte Lichtbewegung lässt sich ferner als ein Zurückgeworfensein deuten, das nicht nur eine sich linear ausbreitende Bewegung visualisiert, sondern auch deren Tendenz, zum Ursprungsort zurückzuweisen. Die Fortleitung von Licht stellt sich als Linie dar, die – wie der Reflex – zurückgeworfen ist. Das Bild könnte somit eine wichtige Quelle für diese ansonsten unklare Bedeutungskomponente liefern, die sich bei Willis physiologisch nicht erschließt. Gerade dieser Bedeutungsaspekt des Reflexes transportiert sich ausschließlich über Bilder, über ihre räumliche Anordnung bzw. Architektur. Auch in anderen Bereichen von Willis'

Abbildung 3: Anonym, Pyrotechnik, 1630.

Gehirnlehre lässt sich ein Argumentieren auf der Basis visueller Darstellung nachweisen. An mehreren Stellen der Schrift *Cerebri anatome* vergleicht Willis das Kleinhirn mit einem Brunnen:

> When some time past I diligently and seriously meditated on the office of the Cerebel [...] it occured, to wit, that the Cerebel is a peculiar Fountain of the animal Spirits designed for some works, and wholly distinct from the Brain. Within the Brain, Imagination, Memory, Discourse, and other more superior Acts of the animal Function are performed; besides, the animal Spirits flow also from the nervous stock, by which all the spontaneous motions, to wit, of which there are knowing and will, are performed. But the office of the Cerebel seems to be for the animal Spirits to supply some Nerves, by which involuntary actions (such as the beating of the Heart, easie Respiration, the Concoction of the Aliment, the pro-

> trusion of the Chyle, and many others) which are made after a constant manner unknown to us, or whether we will or not are performed.[13]

Die Basis dieses Vergleich ist funktional: denn Willis sieht im Kleinhirn (Cerebel) einen Brunnen, der die Nervengeister in die Umlaufbahn der Nerven bringt. Wenn Willis das Kleinhirn mit einem Brunnen vergleicht, basiert dieser Vergleich aber nicht nur auf funktionalen Analogien, also etwa auf der Annahme, dass das Organ einer hydraulischen Pumpe ähnlich ist, welche Animalgeister wie Wasser durch die Nervenbahnen transportiert. Der Vergleich basiert zudem auf visuellen Analogien, die wiederum nur über topographische Ordnungen von Bildern hergestellt werden.

Abbildung 4 zeigt ein Wasserspiel und stammt aus der Schrift Agostino Ramellis *Schatzkammer mechanischer Künste*, die auch René Descartes in seinem *Traité de l'homme* erwähnt.[14] An der Zierkugel des Brunnens, die sich am oberen Bildrand befindet, entspringen sieben Wasserfortleitungen, welche Schlangen darstellen. Abbildung 5 ist eine Hirndarstellung aus Willis' Schrift Cerebri anatome, auf der man den Hirnstamm mit austretenden Hirnnerven erkennt.

Die austretenden Hirnnerven gleichen den Schlangen auf Ramellis Darstellung. Es ist mithin anzunehmen, dass derartige topographische Ähnlichkeiten auch den Vergleich des Hirnstamms bzw. Kleinhirns mit einem Brunnen organisiert haben könnten und den funktionalen Überlegungen beigeordnet sind. Hier wird ein Bild nicht zur Illustration verwendet. Die topographischen Aspekte des Bildes stiften vielmehr die Grundlage für einen Vergleich, der womöglich erst in einem weiteren Schritt zur funktionalen Metapher erweitert wurde. Eine Maschine ist damit zum Bildspender für ein menschliches Organ geworden, die Hydraulik hat bildliches Wissen über das Kleinhirn bereitgestellt.

Bezeichnenderweise folgt auf den Vergleich des Kleinhirns mit einer Maschine sogleich eine Abgrenzung, wie ein Blick auf die oben zitierte Stelle zeigt. Dort wird nicht nur über die Funktions-

---

13) Thomas Willis, *The Anatomy of the Brain and Nerves*, ed. William Feindel, 2 Bde. (Montreal, 1965), Bd. 1, 111.

14) Canguilhem, *Die Herausbildung*, 81.

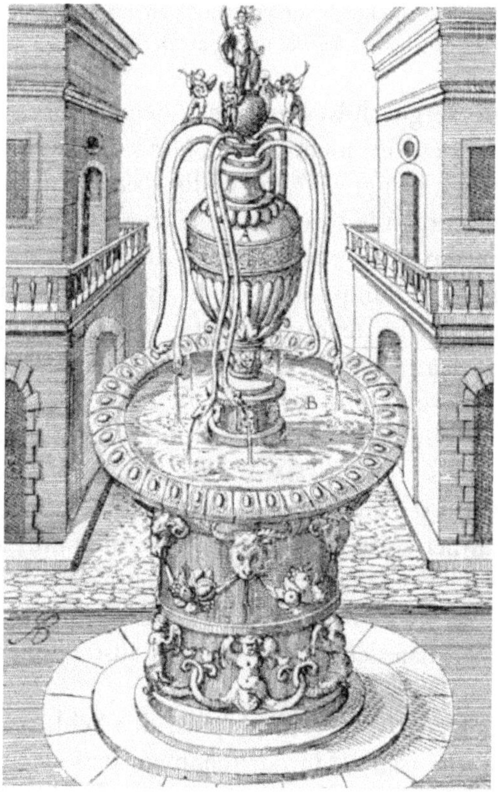

Abbildung 4: Agostino Ramelli, Schatzkammer mechanischer Künste.

weise des Kleinhirns spekuliert. Das Kleinhirn wird zudem vom Großhirn („Brain“) unterschieden und zwar aufgrund anatomischer und psychologischer Überlegungen. Wenn Willis das Kleinhirn als eine Struktur bezeichnet, die vom Großhirn unabhängig ist, dann setzt er eine funktionale und morphologische Distinktheit voraus. Denn er ordnet dem Großhirn bestimmte Vermögen zu, nämlich die Phantasie, das Gedächtnis, die Vernunft, den Willen und den Witz. Diese Lokalisation ist an sich nicht weiter ungewöhnlich. Ihre Koinzidenz mit den technischen Metaphern fällt allerdings ins Auge. Technische Vergleiche provozieren, so könnte man meinen, offenbar derartige Unterscheidungen. Sie erfordern mithin eine Grenzziehung zwischen Großhirn und Kleinhirn und damit eine Unterscheidung

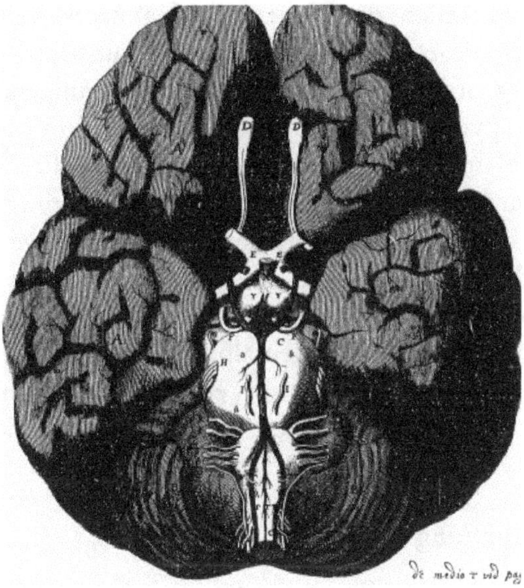

Abbildung 5: Thomas Willis, *Cerebri anatome*, 1680.

zwischen Mensch und Maschine. Die Maschine verfügt weder über ein Großhirn noch über die dort lokalisierten Vermögen Gedächtnis, Wille etc. Die Maschine ist vielmehr, metaphorisch gesprochen, ein Kleinhirn, das die Bewegung von Partikeln, Substanzen oder Flüssigkeiten gewährleistet. Ziehen technische Vergleiche derartige Abgrenzungen nach sich, so führen sie auch zur Annäherung zwischen Mensch und Maschine. Animalgeister, ob als Agenten oder Vermittler von Bewegungen verstanden, befinden sich nämlich sowohl im Kleinhirn als auch im Großhirn, d.h. sie sind – nach Willis' Typisierung – sowohl Träger der willentlichen als auch der unwillentlichen Bewegung. Es stellt sich also die Frage, ob die physiologischen Prozesse, die in Groß- und Kleinhirn ablaufen, in der Tat so verschieden sind. Der Vergleich von Kleinhirn und Brunnen erforderte daher, eine anatomische Grenze zwischen Mensch und Maschine zu ziehen. Diese Grenze verläuft aber nun mitten durch den menschlichen Körper. Sie trennt, wenn man so will, Klein- und Großhirn. Physiologisch gesprochen, ist diese Grenze aber nicht annähernd so stabil wie die anatomische. Denn in den jeweils unterschiedenen Teilen des Gehirns laufen nach Willis dieselben physiologischen Prozesse ab, und dieselben Substrate, die Animalgeister, sind Agenten und Vermittler von Bewegungen.

Betonen die pyrotechnischen Vergleiche und die Visualisierungen das Instantane und das Zurückgeworfene des Reflexes, vermögen sie das Nacheinander von Empfindung und Bewegung nicht angemessen zu erfassen. Die beiden Reflexkomponenten werden auf den Bildern nicht als diskrete, aufeinander folgende Ereignisse fassbar. Ihre Diskretheit scheint vielmehr zugunsten des Instantanen verloren zu gehen. Erst mit den Experimentalpraktiken des 18. Jahrhunderts – so die weitere These – etabliert sich ein anders akzentuiertes Nacheinander, das schließlich physiologisches Reflexwissen komplettiert.

## 2. Die Verzeitlichung

Werfen wir vor diesem Hintergrund – und damit komme ich zum zweiten Teil – einen Blick auf die Reizexperimente des 18. Jahrhunderts, auch auf solche, die in der Regel nicht mit dem Reflex in Verbindung gebracht werden. Dazu zählen etwa die Darmreizexperimente oder die Köpfungsexperimente. In seinem *An Essay on the Vital and Other Involuntary Motions of Animals* von 1751 beschreibt der schottische Arzt Robert Whytt derartige Experimente ausführlich. Er interessiert sich in breitem Maß für Effekte und Stimuli. An die Stelle von Willis' hydro- und pyrotechnischen Metaphern treten nun verschiedenste Experimentalbeschreibungen, die komplexe Manipulationen an Körpern umfassen: etwa das Stimulieren der Nerven durch Duftstoffe, das Erhitzen und Berühren von Körperteilen oder die Entfernung von Nerven bzw. einzelnen Organen.[15]

Dass der Reiz für Whytt – als manipulativer Eingriff – eine zentrale Bedeutung hat, dürfte einerseits mit der Karriere des Experimentes im 18. Jahrhundert zusammenhängen. In Experimenten etablierten sich andererseits Reiz-Reaktions-Abfolgen, womit zunächst – ganz konkret gedacht – Handlungsvollzüge gemeint sind. Diese setzen

---

15) Robert Whytt, *An Essay on the Vital and Other Involuntary Motions of Animals* (Edinburgh, 1751), 83f.

praktisches Wissen voraus, ein know-how. Das Reizen war im 18. Jahrhundert eine Praktik, die an sich variiert und trainiert wurde und auf deren Beherrschung sich die akademischen Bemühungen richteten. Reizversuche waren also nicht nur Mittel zum Zweck, sie wurden selbst Wissensgegenstände und dienten nicht nur der Überprüfung bestimmter Hypothesen.[16] Ein aus dem Umfeld Albrecht von Hallers stammendes Versuchsprotokoll gibt darüber Aufschlüsse:

> Um 11 Uhr stach ich eine Nadel durch das Gehirn eines Frosches. Der Frosch krümmte sich, wenn auch nicht zu stark. Nachdem ich das Gehirn entfernt hatte wurde der ganze Körper des Frosches mehrmals von Konvulsionen erfasst. 15 Minuten danach stieß der Frosch ein lautes Geräusch aus, um 11.25 machte er drei große Sprünge. Um 12.40 warf ich das Tier, das auf dem Boden lag, auf den Rücken. Es drehte sich in seine Ursprungslage um und machte ein paar weitere Sprünge. Um 1.46, als ich den Frosch mit einem Finger reizte, versuchte er ein paar große Sprünge zu machen. Um 2.43 ließ er sich sehr stark von Reizen an den Füßen erregen. Um 3.13 reizte ich ihn mit kaltem Wasser. Der Frosch machte Sprünge, eher größere. Danach legte ich den Frosch ins Wasser – er wurde ganz steif.[17]

Der Experimentator notiert hier einzelne Reiz-Experimente. Wie die Notizen zeigen, wurden sie oft wiederholt, ohne dass dabei ein konkretes Erkenntnisinteresse formuliert wurde. Die Praktiken installierten und automatisieren vielmehr Abläufe. Sie waren Übungen, deren Mechanik kaum zu übersehen ist und die ein regelrechtes Reiz-Regime ausbildeten. Auf diesem Weg vermittelte sich ein praktisches Wissens, das für die späteren Reflexexperimente wichtig wurde. In Experimenten wurden wichtige Beobachtungen gemacht. Die Experimente erlaubten zudem, die von Willis noch als instantan gedachte Empfindungs-Bewegungs-Kaskade in eine temporale Ordnung zu integrieren und damit einen deutlich anderen Akzent

---

16) Hubert Steinke, *Irritating experiments. Haller's concept and the European controversy on irritability and sensibility 1750–90* (Amsterdam, New York 2005); Margarete Vöhringer: „Hallers Köpfungen" in *‚Allerhand nützliche Versuche'. Empirische Wissenskultur in Halle und Göttingen*, ed. Tanja van Hoorn, Yvonne Wübben (Hannover, 2009), 105-121.

17) Johann Georg Zimmermann, *Mit Skalpell und Federkiel. Ein Lesebuch*, ed. Andrea Langenbacher (Bern, 1995), 28.

auf das Nacheinander zu setzen. Ohne diese, erst in den Versuchspraktiken etablierte Form des Nacheinander, hätte sich kein Wissen vom Reflex herausgebildet.[18] Zugleich warfen die Tierexperimente die Frage nach der Vergleichbarkeit von Mensch und Tier auf, die in einem zweiten Schritt zum Transfer von tierexperimentellem Wissen auf Menschen führte, wie sich an Marshall Hall zeigen lässt.

## 3. Anthropotechnische Mischorganismen

Das praktische know-how des Experimentes, und damit komme ich zum dritten und letzten Aspekt, wird für den englischen Physiologen Marshall Hall theoretisch relevant und geht in seine Definition des Reflexes ein. Die eingangs zitierte Passage lautete:

> This [new] property is characterized by being excited in its action, and reflex in its course: in every instance in which it is exerted, an impression made upon the extremities of certain nerves is conveyed to the medulla oblongata or the medulla spinalis, and is reflected along other nerves.[19]

Deutlich verweist „excited“ auf die Versuchsanordnungen des 18. Jahrhunderts. Gemeint ist jetzt keine natürliche Empfindung mehr, also eine zufällige *sensatio* wie das Jucken. Gemeint ist nun ein künstlich erzeugtes Ereignis, eine Reizung, deren Effekt gezielt hergestellt und zeitlich erfasst wurde. Die frühneuzeitliche Instantanität von Empfindung und Bewegung, die noch bei Willis im Vordergrund stand, geht nun verloren bzw. wird in ein zeitlich akzentuiertes Nacheinander transformiert. Hall übersetzt die Instantanität in eine Folge von Reiz (*excited*) und Bewegung. Sie ist keine Eigenschaft des Initialereignisses – der Explosion – mehr, sondern der Reaktion. Dieses Nacheinander, das die Praktiken des 18. Jahrhunderts etablierten, wird bei Hall jetzt auch theoretisch profiliert. Seit Bell/Magendies Forschung sind die anatomischen Strukturen der

---

18) Ähnlich wie die Maschinenvergleiche werfen die Tierexperimente, in denen das praktische Reflexwissen vermittelt wird, die Frage nach der Übertragbarkeit der Ergebnisse von Tier auf Menschen und damit die Frage nach Gemeinsamkeiten unter Unterschieden auf.

19) Marshall Hall, „On the Reflex Function“, 652.

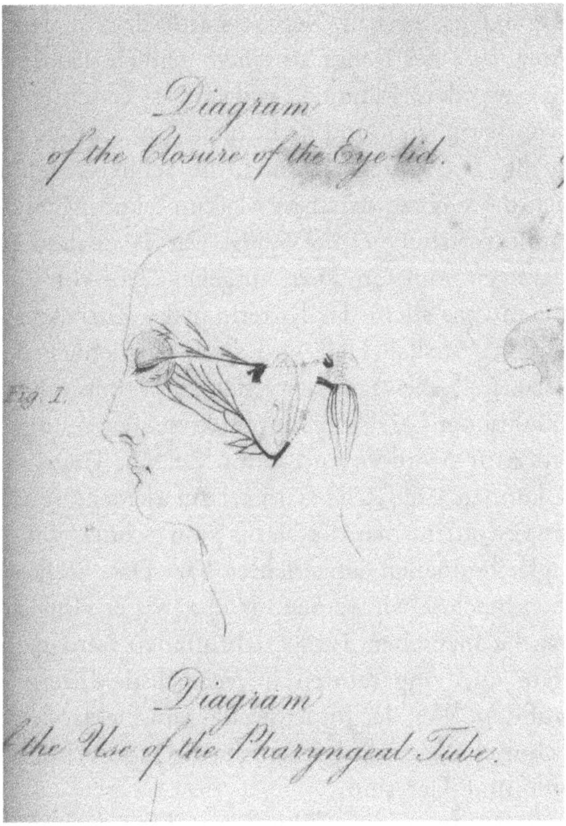


Abbildung 6: Marshall Hall, Reflexbogen Lidreflex, 1841.

Rückenmarkswurzel beschrieben.[20] Es ist bekannt, dass sie aus einem afferenten sensorischen und einem efferenten motorischen Nerven bestehen. Diese Strukturen ordnet Hall dem Reflex zu, den er als ein Rückbiegen entlang von anatomischen Strukturen konzipiert und phänomenologisch zugleich als Nacheinander von Reiz und Bewegung erfasst.

Abbildung 6 stellt die Visualisierung eines Lidreflexes dar. Das Bild zeigt die anatomischen Strukturen, die sensorischen Nerven, die von der Hornhaut zu einem Nervenknoten führen, wo eine

[20] Charles Bell, *Idea of a New Anatomy of the Brain* (London, 1811).

Umschaltung auf motorische Nerven stattfindet. In dem hier visualisierten Ablauf ist der Reflex zu einem zeitlich geordneten, linearen Ereignis geworden. Räumlich gedacht, ist er ferner exzentrisch. Er nimmt von peripheren und nicht mehr von zentralen Strukturen seinen Ausgang. Die Unterscheidung, die Willis vorgenommen hat und die auch die Grenze zwischen Maschine und Mensch festlegte, hat sich somit verschoben. Bei Willis war es noch möglich, dass Reflexbewegungen vom Großhirn ausgehen. Um von einem Reflex zu sprechen, musste allein das Kriterium des Zurückgeworfenseins erfüllt sein. Von Marshall Hall wird diese Möglichkeit definitorisch ausgeschlossen. Reflexbewegungen verlaufen bei ihm unabhängig von zentralen Funktionen. Sie sind Empfindungs-Bewegungs-Kaskaden, die kein Großhirn erfordern und somit Mensch, Tier und Maschine zukommen können. Der Reflex ist jetzt zur elementaren Einheit des Organismus geworden, den die Natur zum Schutz von Tieren und Menschen gleichermaßen eingerichtet hat. Dass Reflexfunktionen bei Menschen beschrieben werden, ist in gewisser Hinsicht als Folge der Tierversuche anzusehen. Im 19. Jahrhundert werden zudem technische Geräte und Apparaturen in Versuchstiere verpflanzt. Z.B. entstehen auf dem Weg der Implantation von Elektroden in Frösche anthropotechnische Modellorganismen, die regelrechte Mischwesen aus Maschine und Tier sind.[21]

Dass der Reflex schon bei Willis ein anthropotechnischer Begriff ist, hat bereits Canguilhem angemerkt, indem er auf die pyrotechnischen Kontexte verwies. Die historischen Bedeutungsverschiebungen erschließen sich jedoch erst, wenn man Bilder und Praktiken des Reflexes angemessen, d.h. in ihren eigenen epistemischen Leistungen einbezieht und sie nicht als Mittel zum Zweck versteht. Willis' Reflex basiert auf räumlichen Analogien, die sich erst durch die Konsultation von visuellen Darstellungen aus der Pyrotechnik ermitteln ließen. Ohne die Explosionsbilder der Frühen Neuzeit hätte Willis die Idee des Zurückgeworfenseins womöglich nicht in

[21] Hans-Jörg Rheinberger, „Überlegungen zum Begriff des Modellorganismus in der biologischen und medizinischen Forschung" in *Modelle des Denkens. Streitgespräch in der Wissenschaftlichen Sitzung der Versammlung der Berlin-Brandenburgischen Akademie der Wissenschaften am 12. Dezember 2003*, ed. Sonja Ginnow (Berlin, 2005), 69-74.

die Definition von *motus reflexus* einbezogen worden. Ohne die Praktiken des 18. Jahrhunderts wäre aus dem instantanen Modell kein zeitlich geordnetes Nacheinander geworden. Denn in den Experimentalpraktiken des 18. Jahrhunderts, so bei Whytt, wird das Nacheinander als Handlungswissen installiert. Die Praktiken bringen Wissen hervor, das auch im Reflexexperiment Anwendung findet und den Reflex handhabbar bzw. konzeptualisierbar macht. Dieser Aspekt des Reflexwissens vermittelt sich nicht über Begriffe und Texte, sondern primär über Experimentalpraktiken. Erst so wird klar, wie Hall einen Reflexbegriff formulieren konnte, indem er nämlich über praktisches Wissen verfügte, über die Vorstellung einer zeitlichen Folge sowie über die des Zurückwerfens, die nun auf anatomische Strukturen übertragen wird. Hall hat Bild und Praktik damit auf den Begriff gebracht.

Über die angemessene Einbeziehung von Bildern und Praktiken eröffnet sich auch die anthropotechnische Dimension des Reflexes. Die visuellen Analogien sind dem Bereich der Pyrotechnik entnommen. Sie erforderten zunächst Abgrenzungen, d.h. eine Unterscheidung zwischen Maschine und Mensch, die sich im Blick auf physiologische Vorgänge zuweilen jedoch als instabil erwies. Über die Tierversuche des 18. Jahrhunderts etablierten sich schließlich Praktiken, die im Verlauf des 19. Jahrhunderts zur Erzeugung jener anthropotechnischen Organismen führen, die Mischwesen aus Natur und Technik sind und die Unterscheidbarkeit zwischen Mensch, Tier und Maschine letztlich in Frage stellen.

# Machina Machinarum.
# Die Uhr als Begriff und Metapher zwischen 1450 und 1750

**Hanns-Peter Neumann**
*Martin-Luther-Universität Halle-Wittenberg**

**Abstract**
The changing use of the clock metaphor serves as a helpful contrast medium to highlight the different concepts of the body-soul-system between 1450 and 1750. This article first relates to the social, political and philosophical functions of the *horologium*. Then it outlines the different fields of discourse, in which the clock metaphor was mainly invoked. Finally, it examines the writings of a number of significant authors (Cusanus, Ficino, Descartes, Leibniz, Wolff, La Mettrie) with an eye to the evolution of the clock metaphor in various theological, metaphysical and physiological contexts. Surprisingly enough, the clock (or watch) initially represented the life-giving soul and human consciousness, before turning into the well-known symbol for the body-machine, and in particular for its neurophysiological operations.

**Keywords**
clock metaphor, automaton, temperance, self-regulation, body-soul-system, consciousness, cosmology, (neuro)physiology

* Martin-Luther-Universität Halle-Wittenberg, Interdisciplinary Centre for European Enlightenment Studies, Franckeplatz 1, Haus 54, 06099 Halle (Saale), Germany (hanns-peter.neumann@izea.uni-halle.de).

What Skill is in the frame of Insects shown?
How fine the Threds, in their small Textures spun?
How close those Instruments and Engines knit,
Which Motion, and their slender Sense transmit?
Like living Watches, each of these conceals
A thousand Springs of Life, and moving Wheels.
Each Ligature a Lab'rynth seems, each part
All wonder is, all Workmanship and Art.
– *Richard Leigh (1675)*

... que les horloges s'accorderoient plustot que
les philosophes.
– *Leibniz*

## 1. Heuristische Indikamente des Uhr-Modells

In ihrem Standardwerk zur Bedeutung organismischer Metaphern, *Les métaphores de l'organisme*, geht Judith Schlanger auch kurz auf die Uhr als logisches Modell für das mechanistische Verständnis der Welt und der Gesellschaft ein. Sie interpretiert das Uhr-Modell als „un concept purement métaphysique"[1] und skizziert einige grundlegende Bestimmungen, die den Begriff der Uhr auszeichnen:

> So ist die Uhr wahrhaftig die Maschine der Maschinen, d. h. sie kann offensichtlich genauso als die autonomste wie heteronomste aller Maschinen betrachtet werden. Sie ist schlechterdings die Maschine, die von sich aus funktioniert, das geschlossene und belebte, mit einer inneren (Selbst)Regulation ausgestattete System. In diesem Sinne ist sie gleichfalls der Mechanismus, der andauert, das Objekt, das niemals in der Zeit untätig bleibt, sondern die Zeit dadurch präzise meistert und in Besitz nimmt, dass sie sie misst.[2]

---

1) Judith Schlanger, *Les métaphores de l'organisme* (Paris, 1995; 1. Auflage Paris, 1971), 52.

2) Schlanger, *Les métaphores*, 53: "Si l'horloge est véritablement la *machina machinarum*, c'est qu'elle peut être considérée avec autant d'évidence comme la plus autonome et comme la plus hétéronome des machines. Elle est par excellence la machine qui fonctionne seule, le système clos et animé, doué de régulation interne. Elle est également, en ce sens, le mécanisme qui dure, l'objet qui ne demeure pas inerte dans le temps, mais qui précisément maîtrise et possède le temps, puisqu'il le mesure." – Die Übersetzungen aus dem Französischen und dem Lateinischen stammen hier und im Folgenden, soweit nicht anders angegeben, vom Verfasser dieses Beitrags.

Schlanger deutet die Uhr als metaphysische Idee der Maschine, als *machina machinarum*, die allen besonderen Maschinen ideell zugrunde liegt. Als autonomes, geschlossenes und „belebtes" System, das sich selbst reguliert, repräsentiert die Uhr die auf Permanenz ausgelegte Zweckmäßigkeit mechanischer und organismischer Gebilde.

Folgt man Judith Schlangers Definition der Uhr als Maschine schlechthin, so ist diese deswegen ein mehr oder weniger präzises Regulativ der Zeit, weil sie das *tempus*, das als konkretes Verhältnis des Menschen zu Tag und Nacht zunächst nur qualitativ (hell-dunkel) und empirisch-ereignishaft (Sonnenaufgang, Sonnenuntergang; Tagesanbruch, Anbruch der Dunkelheit) bestimmt ist, nun auf abstrakte Weise nach arithmetischen Zähleinheiten misst und, unabhängig von Lichtverhältnissen, meteorologischen Verhältnissen und Tageszeiten, meistert (*maîtrise*) und in Besitz nimmt (*possède*).[3] Neben ihrer Funktion der maschinellen und mechanischen Fundierung der Zeit als abstraktes Zählmaß für Dauerhaftigkeit und Sukzession – in ihr als in strikter Taktung und Rhythmisierung fortlaufende Bewegung (*perpetuum mobile*) realisiert, die sich in allen Maschinen und organisierten Körpern, besonders in deren Kreisläufigkeit, ausdrückt – hat die Uhr aber noch die nur ihr zukommende spezifische und alltägliche Aufgabe der Zeitindikation. Als solche dirigiert, terminiert und reguliert sie das soziale Leben nachhaltig.

Doch genausowenig wie die Uhr immer schon ein ahistorisches metaphyisches Konzept war und ist, kann die heute konventionelle Form der zeitlichen Organisation des sozialen Lebens nach einem abstraktem, horologisch realisiertem Zählmaß Anspruch auf Zeitlosigkeit erheben. Dieser in den Ohren jedes Historikers sicher banal klingende Hinweis scheint mir angesichts der von Schlanger vorgenommenen metaphysischen Konzeptualisierung der *horloge* allerdings nicht unangebracht.

---

3) Vgl. dazu die Definition der Uhr bei John Smith, *Horological Dialogues. In Three Parts. Shewing The Nature, Use, and Managing of Clocks and Watches […]* (London, 1675), 3: "A Clock is a certain instrument invented and contrived, to give the true hour, either of day or night, (by the Circuler moving or certain wheels and pinions artificially disposed within the Body of the Instrument) without the help of either Sun, Moon, or Stars, or any other of those elemental assistances, which the Antients [sic!] did in former ages make use of to divide their time by."

Dennoch sind viele der von Schlanger im obigen Zitat genannten Charakteristika in der Tat wichtige Indikamente für den Gebrauch der Uhr als Begriff und Metapher im Zeitraum zwischen 1450 und 1750.[4]

Im Folgenden möchte ich daher Schlangers „rein metaphysisches Konzept“ (“concept purement métaphysique”) als heuristisches Modell verwenden, das hilfreiche Indikatoren zur Bestimmung der Funktion der Uhr als Begriff und Metapher in der frühen Neuzeit und in der Aufklärung zur Verfügung stellt. Zu diesen Indikatoren zählen (1) das in sich geschlossene, belebte System; (2) die systemimmanente Selbstregulation; (3) die permanente Tätigkeit und Bewegung; (4) die Bändigung und Inbesitznahme der Zeit durch (5) Maß und Messen.

Zunächst wird es um (1) und (3), um die Uhr als in sich geschlossenes, belebtes, in permanenter Bewegung befindliches System gehen, schließlich um (2), (4) und (5), also um die Uhr als selbstregulativer Automat, der die Zeit durch Maß und Messen bändigt und vereinnahmt. Dabei ist es vor allem das Moment der – freilich im Laufe der historischen Entwicklung der Uhr nicht immer schon vollkommen ausgebildeten und präzisen – Selbstregulation der Uhr, das sowohl für die permanente Bewegung als auch für die horologische Zeitbändigung konstitutiv ist.

### *1.1 Die Uhr als geschlossenes System permanenter Tätigkeit und Bewegung*

Den Indikamenten des geschlossenen, belebten Systems und der permanenten Tätigkeit und Bewegung entspricht weitgehend der frühneuzeitliche, an antiken Vorgaben reflektierte Begriff des Automaten, für das das *horologium* als musterhaftes Beispiel dient.[5] Dabei sprach man den Automaten aber keineswegs Leben im buchstäblichen Sinne zu, sondern nur den *Anschein* des Lebendigen. So schreibt Giovanni Tortelli 1484: „Von daher scheint es [das Horologium] gewissermaßen Leben zu haben, weil es sich von sich aus und aus eigenem Antrieb [*sua*

---

4) Für einen ersten instruktiven Überblick über die Geschichte der Uhr-Metapher vgl. *The Encyclopedia of Time*, Hg. Samuel L. Macey (New York, 1994), 113 ff.

5) Zur Rolle des Automaten in der Technikgeschichte der frühen Neuzeit vgl. Silvio Bedini, “The Role of Automata in the History of Technology,” *Technology and Culture*, 5 (1964), 24-42. Vgl. zudem neuestens Jonathan Sawday, *Engines of the Imagination. Renaissance culture and the rise of the machine* (London/New York, 2007).

*sponte*] regt und tönt sowie Tag und Nacht sein Werk für den Menschen verrichtet."[6] Der Florentiner Renaissancehumanist Angelo Poliziano charakterisiert die Uhr im Kapitel 97 seiner *Centuria Miscellaneorum* um 1488/1489 wie folgt:

> Offenbar pflegt man also im eigentlichen Sinne Automaten jene zu nennen, welche Mechaniker so gebaut haben, dass sie sich auf irgendeine Weise wie aus einer uns unbekannten Ursache von sich aus und aus eigenem Antrieb [*sua sponte*] in Tätigkeit zu setzen scheinen. Von solcher Art sind heute entweder die Horologia oder die auf irgendeine Weise sich drehenden Maschinen, an denen wir Bildchen bewundern, die durch eine verborgene Kraft [*occulta vi*] verspielt umherlaufen.[7]

Polizianos Verweis auf die *mechanici* kann möglicherweise bereits als Indiz für die frühneuzeitliche Aufwertung der antiken Tradition der mechanischen Künste (etwa eines Heron von Alexandria) gelesen werden.[8] Wichtig ist allemal, dass Poliziano das Horologium mit dem Automatenspiel bewegter Figuren zusammendenkt und von „sich drehenden Maschinen" ("versatiles machinae") spricht.[9] Letzteres deutet

---

6) Giovanni Tortelli, *De orthographia dictionum e Graecis tractarum* (Venedig, 1484); hier zitiert nach der Edition des Horologium-Artikels aus *De ortographia* in: Alex Keller, "A Renaissance Humanist Looks at 'New' Inventions: The Article 'Horologium' in Giovanni Tortelli's *De orthographia*," *Technology and Culture*, 11 (1970), 349: "Unde quodammodo vitam habere videtur: cum sponte sua cietur: & dies ac noctes pro homine opus facit: quo nihil neque utilius neque iucundius dici potest: novum tamen inventum est."

7) Zitiert nach folgender Ausgabe: Angelo Poliziano, *Miscellaneorum Centuria prima*, in *Omnium Angeli Politiani Operum tomus prior* (Venedig, 1512), Fol. CXLII^v: "Apparet ergo Automata appellari solita, quae ita mechanici fabricabantur ut sua sponte efficere quippiam, uelut ignara causa uiderentur. Qualia nunc aut horologia sunt, aut versatiles quaepiam machinae, in quibus imagunculas, occulta vi, cursitantes, ludibundasque miramur."

8) Zur Reklassifizierung der *artes mechanicae* im Gesamtgefüge der *artes liberales* in der frühen Neuzeit vgl. Helen Hattab, "From Mechanics to Mechanism," in *The Science of Nature in the Seventeenth Century. Patterns of Change in Early Modern Natural Philosophy*, Hg. Peter R. Anstey u. John A. Schuster (Dordrecht, 2005), 99-129, insbes. 100-116.

9) Poliziano hatte bereits 1484 die seinerzeit berühmte astronomische Uhr Lorenzos della Volpaia lobend erwähnt und diesen selbst als *Archimedes redivivus* bezeichnet. Vgl. dazu Stéphane Toussaint, "Ficino, Archimedes and the Celestial Arts," in *Marsilio Ficino: His Theology, His Philosophy, His Legacy*, Hg. Michael J. B. Allen, Valery

darauf hin, dass sich schon im späten Quattrocento der Begriff der *machina* in seiner mittelalterlichen Bedeutung als stabiles statisches (Bau)Gerüst (oder als meist hölzerne Stützkonstruktion) hin zur Konnotation der Maschine als dynamisches, selbstbewegendes und in sich stabiles Gebilde geöffnet hatte.[10]

Auch Polydorus Virgilius aus Urbino preist im letzten Kapitel des dritten Buches, also an exponierter Stelle seines 1499 in Venedig publizierten, seinerzeit zu Berühmtheit gelangten Werkes *De rerum inventoribus libri tres* die Erfindung der Uhr, indem er auf deren selbsttätigen Stundenschlag hinweist: „Denn was konnte Erfreulicheres erfunden werden als das Horologium, durch das uns mittels einer, so wie es scheint, von sich aus und aus eigenem Antrieb [*sua sponte*] tönenden Glocke die Stunden mitgeteilt werden, selbst dann, wenn die Sonne verborgen ist?"[11]

In allen genannten Fällen erweist sich das „sua sponte" als signifikantes, mit und in der Uhr mechanisch realisiertes Merkmal der vitalisierenden Funktion der *anima*, die im Quattrocento für das den Körper regulierende Prinzip der Selbstbewegung und des Lebens schlechthin gehalten wurde. Dabei ist man sich, wie gesagt, des bloßen Augenscheins (*videri*) der Lebendigkeit des horologischen, stundenschlagenden Automaten durchaus bewusst gewesen. Von grundlegender

---

Rees u. Martin Davies (Leiden, 2002), 307-326; zu Poliziano s. ebd., 307-308. Zu Lorenzo della Volpaia und seiner Planetenuhr vgl. Umberto Dorini, "L'orologio dei pianeti di Lorenzo della Volpaia," *Rivista d'arte*, 6 (1909), 137-144.

10) Vgl. dazu Marcus Popplow, „Die Verwendung von lat. *Machina* im Mittelalter und in der Frühen Neuzeit – vom Baugerüst zu Zoncas mechanischem Bratenwender," *Technikgeschichte*, 60 (1993), 7-26. Zur Bedeutungsgeschichte des *machina mundi*-Topos im Mittelalter und in der frühen Neuzeit vgl. Marcus Popplow, "Setting the World Machine in Motion: The Meaning of *Machina Mundi* in the Middle Ages and the Early Modern Period," in *Mechanics and Cosmology in the Medieval and Early Modern Period*, Hg. Massimo Bucciantini, Michele Camerota u. Sophie Roux (Florenz, 2003), 45-70.

11) Polydore Vergil, *De inventoribus – On Discovery*, hg. und übersetzt von Brian P. Copenhaver (Cambridge, Mass./London, 2002), 484: "Nam quid iucundius reperiri potuit horologio, quo nobis etsi occultata sole per tintinabulum sua ut ita videtur sponte sonans horae nuntiantur?"

Bedeutung aber war die technische Fertigkeit der *mechanici*, die Natur in ihrer Vitalität nachahmen bzw. simulieren zu können.[12]

Die Uhr – *horologium*, *horloge*, *montre*, *clock*, *watch*, später auch *pendule* und *pendulum* genannt – war aber zusätzlich noch dadurch ein besonders herausragendes Sinnbild der vitalen, okkulten Selbstbewegungskräfte der Natur, dass sie als „Astrarium"[13] und als Zeitindikator die perfekte Bewegung schlechthin, nämlich die Kreisbewegung, hier vor allem die Kreisbewegung der Himmelskörper, mechanisch ab- und nachbildete. Sie symbolisierte ein bei aller Bewegung, die als *mutatio* gemeinhin für geringer geachtet wurde als die Ruhe des in sich gleichbleibend Verharrenden, geschlossenes, in seinem Kreislauf und seiner *rotunditas* aber quasi schon ruhendes, wohlorganisiertes Universum.

### *1.2 Die Uhr als selbstregulativer Automat der Zeitmessung und Zeitbändigung*

Neben den Indikatoren des geschlossenen, belebten Systems und der permanenten Tätigkeit und Bewegung haben auch die Momente der systemimmanenten Selbstregulation, der Bändigung und Inbesitznahme der Zeit durch Maß und Messen ihre historische Entsprechung im Spätmittelalter. Das mechanische Stundenschlagwerk und die mechanische Uhrwerkhemmung, die vermutlich in Italien im Zeitraum

---

12) Die auffällige Übereinstimmung der verschiedenen Beschreibungen lässt sich noch durch eine frühere Quelle ergänzen. So wurde die erste Turmuhr Moskaus (1404) wie folgt beschrieben: „Dieser Stundenzeiger heißt Stunden-Messer; zu jeder Stunde schlägt er einen Hammer auf eine Glocke, zählend und messend die Stunden des Tages und der Nacht [...] Niemand schlägt sie, es ist vielmehr wunderbar und merkwürdig anzusehen, *wie etwas sich auf menschliche Weise von selbst bewegt und erklingt durch menschliche Fertigkeit mit großer Erfindungsgabe und Schlauheit.*" (Hervorheb. HPN), zitiert nach Gerhard Dohrn-van Rossum, *Die Geschichte der Uhr. Uhren und moderne Zeitrechnungen* (Köln 2007; 1. Ausgabe München 1992), 148.

13) Das zwischen 1348 und 1364 entworfene und konstruierte Astrarium war die seinerzeit berühmte astronomische Uhr Giovanni de' Dondis, vgl. dazu Silvio A. Bedini u. Francis R. Maddison, "Mechanical Universe: The Astrarium of Giovanni de' Dondi," *Transactions of the American Philosophical Society*, 56 (1966), 1-69, vgl. insbes. S. 50-54 die Quellendokumente, die die Rezeption des Astrarium bei den zeitgenössischen Gelehrten belegen. Eine ähnliche Berühmtheit erlangte die astronomische Uhr von Richard von Wallingford aus Oxford (1292-1336), vgl. dazu John North, *Richard of Wallingford and the Invention of Time* (London/NewYork, 2005).

zwischen 1280 und 1330 entwickelt wurde und zur Zeit Giovanni Tortellis, Angelo Polizianos und Polydore Vergils bereits weit verbreitet war, machten es zum einen möglich, den Glockenschlag der Uhr mechanisch so zu regulieren, dass die jeweilige Stunde durch die Anzahl der Schläge signalisiert wurde,[14] zum andern, dass das Uhrwerk gleichmäßig ablief und somit die Sequenzen zwischen den Zählstunden in gleicher Länge abgemessen werden konnten. Die Präzisionsdauer ließ allerdings noch lange Zeit zu wünschen übrig. Vor allem die Stundenschlaguhr erzeugte großes Aufsehen und war, wie die Quellen belegen, eine spätmittelalterliche und frühneuzeitliche Sensation, während die Uhrwerkhemmung selbst zunächst weniger wahrgenommen wurde.[15] Im Laufe des 15. Jahrhunderts feierten die öffentliche, aber auch die häusliche, ja sogar die tragbare Uhr schließlich Konjunktur.[16] So wurde die öffentliche städtische Turmuhr zu einem Prestigeobjekt von erheblicher sozialer und politischer Tragweite.[17] Sie war nicht nur dazu bestimmt, die Bürger zu einem geordneten und tugendhaften Leben aufzurufen. Auch deren Zustand, Pflege und Wartung waren Kriterien der ländlichen und kirchlichen *Policey*, die Integrität der Städte und der städtischen Verwaltung zu prüfen und zu kontrollieren.[18] Noch 1680 verweist der Physiker, Alchemiker und Merkantilist Johann

---

14) Das von Tortelli verwandte Verb *ciere* in "sua sponte cietur" hat auch die Bedeutung von *ertönen lassen*, vgl. das Zitat in Anm. 6.

15) Vgl. dazu und zu den folgenden Ausführungen die grundlegende Studie von Dohrn-van Rossum, *Geschichte der Uhr*, 68-76 und v. a. 138-151. Siehe weiterhin Gustav Bilfinger, *Mittelalterliche Horen und Moderne Stunden* (Wiesbaden, 1969), auf dem Dohrn-van Rossum kritisch aufbaut, sowie die summarische Darstellung bei Carlo M. Cipolla, *Gezählte Zeit. Wie die mechanische Uhr das Leben veränderte* (Berlin, 1999; 1. Ausgabe unter dem Titel *Le macchine del tempo* Bologna, 1981). Vgl. a. David S. Landes, *Revolution in Time. Clocks and the making of the modern world* (London, 2000, revidierte Ausabe der Erstedition 1983).

16) Die Bezeichnung *horologium publicum* findet sich zwar schon bei Petrarca, wird aber erst ab dem Ende des 15. Jahrhunderrts regelmäßig gebraucht, vgl. Dohrn-van Rossum, *Geschichte der Uhr*, 170.

17) Vgl. dazu den Überblick bei Sigrid u. Klaus Maurice, „Stundenangaben im Gemeinwesen des 16. und 17. Jahrhunderts," in *Die Welt als Uhr. Deutsche Uhren und Automaten 1550-1650*, Hg. Klaus Maurice u. Otto Mayr (München/Berlin, 1980), 146-158.

18) Vgl. Dohrn-van Rossum, *Geschichte der Uhr*, 181, 196 ff. Die öffentlichen Uhren mussten regelmäßig, wenn nicht gar ständig gewartet und neu justiert werden, da sie, je nach Qualität oder Materialermüdung, genauer oder ungenauer gingen. Mitunter

Joachim Becher in einem an die Royal Society gerichteten Traktat *De nova temporis dimetiendi ratione* (London, 1680) auf den allgemeinen Nutzen der Uhr sowie auf ein deutsches Sprichwort, das die oben genannte sozialpolitische Funktion der Uhr im öffentlichen Leben widerspiegelt:

> Die Methode der Zeitmessung [...] war somit wegen ihres unbedingten und großen Nutzens in den astronomischen Disziplinen und beim Herausfinden der Längengrade sowie im bürgerlichen und öffentlichen Bereich (von daher das deutsche Sprichwort, äußeres Merkmal eines gut eingerichteten Gemeinwesens sei es, dass die Turmuhr akkurat laufe) immer schon überdacht und verbessert worden und von daher immer schon verschieden und vielfältig.[19]

Möglicherweise war es die Uhrwerkhemmung selbst, möglicherweise die dem öffentlichen Bewusstsein wesentlich stärker eingeschriebene sozialpolitisch-regulative Funktion der Uhr, die zur engen Affiliation der Uhr mit der *virtus* der *Temperantia* oder *Attemprance* geführt hat.

Jedenfalls lässt sich vom Spätmittelalter an eine parallel zur Entwicklung der mechanischen Uhr einsetzende zunehmende ikonographische Verbindung der *Temperantia* mit der Uhr (*Temperantia* als *virtus* des Messens und des Maßes), mit der göttlichen *Sapientia* (die sapientiale Ordnung der Welt nach Maß, Zahl und Gewicht im *Liber Sapientiae* des Alten Testaments) und dem mit der Weisheit Gottes typologisierten Logos Jesus Christus konstatieren.[20] Es ist diese auffällige Kohärenz

---

geschah dies, indem man die Gleichmäßigkeit des Uhrwerkgetriebes an normierten Sanduhren überprüfte.

[19] Johann Joachim Becher, *De Nova Temporis Dimetiendi Ratione, & accurata Horologiorum Constructione, Theoria & Experientia. Ad Societatem Regiam Anglicanam in Collegio Greshamensi Londini* (London, 1680), 1-2: "Temporis dimetiendi ratio [...] ita ob necessarium & magnum usum tum in Astronomicis, & gradibus Longitudinis inveniendis, tum in civilibus & publicis (hinc in Germanico idiomate Proverbium: Reipublicae bene constitutae externum signum Horologium in Turri accuratè procedens esse) semper in consideratione & emendatione, ac proinde semper varia & multiplex fuit."

[20] Vgl. dazu die, soweit ich sehe, immer noch einzige ausführlichere Studie zur ikonographischen Entwicklung der *Temperantia* in der frühen Neuzeit von Lynn White Jr., "The Iconography of *Temperantia* and the Virtuousness of Technology," in *Action and Conviction in Early Modern Europe*, Hg. Theodore K. Raab u. Jerrold E. Seigel (Princeton/New Jersey, 1969), 197-219, insbes. 202 ff.

von moralischem Maß (nach dem Vorbild der aristotelischen Ethik), messendem Verstand, göttlicher Weisheit, christlicher Frömmigkeit, bewegungs- und körperregulierendem Prinzip (ein Prinzip, das vor allem in der Astronomie eine große Rolle spielt, so dass man sich immer auch die Frage des Unterschieds von himmlischer und irdischer Physik stellen muss),[21] handwerklicher Fertigkeit und Mechanik, die das metaphorische Potential des Uhr-Modells füttert und dessen diskursübergreifende Variabilität ausmacht.[22]

Über die *Temperantia* und deren regelnde Tätigkeit des Ordnungstiftens konnten die Regierung des politischen Körpers durch die Uhr, die Selbstdisziplin des Menschen, die Uhrwerkhemmung als mechanische *Mäßigung* des Uhrwerks und die vernünftige Regulation des Körpers miteinander analogisiert werden. Eine solche Analogie von Uhrwerkmechanismus, *Temperantia* und menschlichem Körper wiesen schon die astromedizinischen Aspekte öffentlicher Uhren selbst auf, so z. B. die berühmte Straßburger Uhr, erbaut zwischen 1352 und 1354, und die öffentliche Uhr in Mantua.[23] Beide indizierten u. a. die adäquaten Zeiten für medizinisch-wundärztliche Eingriffe (z. B. den Aderlass) und veranschaulichten so die temperierende und vernunftgemäße

---

21) Mindestens seit Nicolaus von Oresme besteht die Vorstellung, dass sich, entgegen der Annahme der Buridanischen *impetus*-Theorie, die zirkuläre Bewegung der himmlischen Sphären beschleunigen würde, gäbe es keine in die himmlische Mechanik integrierte Friktion. Vgl. dazu die kritische Diskussion der keineswegs immer eindeutigen Position Oresmes bei Marshall Clagett, "Nicole Oresme and Medieval Scientific Thought," *Proceedings of the American Philosophical Society*, 108 (1964), 298-309, 300 f.

22) Zu den moralisch-religiösen, affektregelnden Implikationen der moderierenden *Temperantia* vgl. das Lemma "Temperantia" in V. P. Robert, *Aurifodina universalis scientiarum divinarum atque humanorum ex fontibus aureis sanctorum patrum, conciliorum, doctorum, nec non paganorum fere ducentorum tam in theologia, quam in philosophia [...]*, Bd. 4 (Paris, 1868), 252-254. Die in ihrer Transformierbarkeit wohl am weitest gehende Bestimmung von *Temperantia* innerhalb der Zitatensammlung der *Aurifodina* stammt von Albertus Magnus: "Temperantia vera et perfecta, est justa moderatio motuum interiorum, et exteriorum operum." ("Die wahre und vollkommene Temperantia besteht in der richtigen Mäßigung von inneren Bewegungen und äußeren Werken.")

23) Vgl. zur im 16. Jahrhundert neu konstruierten Straßburger Uhr Cunradus Dasypodius, *Warhafftige Außlegung des Astronomischen Uhrwercks zu Straßburg [...]* (Straßburg, 1578).

Funktion der Medizin hinsichtlich der ärztlichen Instandsetzung des kranken Körpers auf eindrückliche Weise.[24]

Besonders deutlich wird die metaphorische Assoziation von Uhrwerkhemmung (oder Uhrwerkwartung) und Regulation des Körpers in Christine de Pizans um 1400 entstandener *Epistre Othea*:

> Die Mäßigung wurde auch Göttin genannt. Und weil nun unser Körper aus verschiedenen Dingen zusammengesetzt ist und in vernünftiger (verhältnismäßiger) Weise gemäßigt werden muss, vermag er durch die Uhr symbolisiert zu werden, die mehrere Räder und Messungen hat; funktioniert die Uhr aber in keinster Weise, wenn sie nicht gemäßigt ist, so kann auch analog dazu unser Körper nicht funktionieren, wenn die Mäßigung ihn nicht reguliert. [25]

Bedenkenswert an diesem Zitat aus Pizans *Epistre Othea* ist vor allem eine dem Vergleich zwischen Uhr und menschlichem Körper inhärente Differenzierung: Dasjenige, was der Uhr ihren Wert gibt, ist das temperierende Moment, das von der Uhrwerkhemmung ausgeht. Diese erwähnt Christine de Pizan allerdings nicht ausdrücklich, so dass es nicht sicher ist, ob sie tatsächlich an die mechanische Hemmvorrichtung des Horologium gedacht hat.[26] Es ist durchaus möglich, dass sie stattdessen auf die übliche technische Justierung und Wartung der öffentlichen Uhren durch *mechanici* anspielen wollte. Auf ähnliche Weise wie die Uhrwerkhemmung oder aber der von der Obrigkeit mit der Wartung beauftragte *mechanicus* der Uhr ihren für ihr öffentliches Funktionieren so wichtigen gleichmäßigen Lauf garantiert,

---

24) Vgl. Nancy Mason Bradbury u. Carolyn P. Colette, "Changing Times: The Mechanical Clock in Late Medieval Literature," *The Chaucer Review*, 43 (2009), 351-375, 358; Cipolla, *Gezählte Zeit*, 41.

25) Christine de Pizan, *Epistre Othea*, Hg. Gabriella Parussa (Genf, 1999), 202: "Attrempance estoit aussi appelle deesse. Et pour ce que nostre corps humain est composé de diverses choses et doit estre attrempé selon raison, peut estre figuré a l'orloge qui a plusieurs roes et mesures; et toutefoiz ne vault rien l'orloge, s'il n'est attrempé, semblablement non fait nostre corps humain, se attrempance ne l'ordonne." – Den Hinweis auf diese Stelle verdanke ich dem instruktiven Aufsatz von Bradbury u. Colette, "Changing Times," 351-375. Hier finden sich auch weitere Beispiele, u. a. aus Chaucers *Canterbury Tales*.

26) Die auf diese Verbindung abzielende Interpretation von Bradbury und Colette ist zwar verlockend, lässt sich aber nicht eindeutig belegen, vgl. Bradbury u. Colette, "Changing Times," 351, 362.

verleiht auch die gleichsam justierende Regulation „selon raison" dem menschlichen Körper erst Wert und Würde. Implizit wird mit der „raison" nun aber das rationale Vermögen der Seele angesprochen, den Körper – über die *spiritus* oder Animalgeister – lenken und koordinieren zu können. Die horologische Maschinenmetapher besitzt im Spätmittelalter und in der frühen Neuzeit daher eine wichtige psychologische Komponente, die es ermöglichte, die Uhr oder die regulierenden Teile der Uhr (Gewichte, Federn, Pendel etc.) mit der den Körper über die *spiritus* vermittelten bewegenden und vitalisierenden Kraft der Seele zu analogisieren. Die öffentliche Uhr konnte somit das hierarchische Verhältnis zwischen Körper und Seele als zugleich technisches und politisches Verhältnis repräsentieren.[27]

Gegen Ende des 17. Jahrhunderts war die regulative Funktion der Uhrwerkhemmung, die im 14. und 15. Jahrhundert noch wesentlich weniger Aufsehen erregt hatte als der Stundenschlagmechanismus, dann aber längst ins öffentliche Bewusstsein vorgedrungen. Das lag nicht zuletzt an der von Christiaan Huygens 1657 konstruierten Pendeluhr, die ein hohes Maß bisher nicht gekannter Präzision in der Zeitmessung bot. So antwortet der *Articus* in der horologischen Schrift *Horological Dialogues* (London, 1675) des englischen Uhrmachers John Smith auf die Frage, woher die Uhr ihre verschiedenen Namen beziehe, u. a. wie folgt: „Secondly, from that, which retards and regulates the violent motion of the wheels, being either, balance or pendulum."[28]

Wenn Leibniz sich dann z. B. mit der Bezeichnung *pendule* oder *pendulum* der Uhr als Metapher bediente, so hatte er die arithmetisch formulierbare Dynamik der regulativen, das Uhrwerk ("the violent motion of the wheels") zur Präzision nötigenden Uhrwerkhemmung im Sinn und konnte damit, wie ich noch zeigen werde, sowohl die Seelenmonade und den Körper mit dem horologischen Automaten versinnbildlichen.

---

27) Zur wichtigen Verbindung der Körper-Maschine-Metapher mit politischer Terminologie in der frühen Neuzeit vgl. die kurze allgemeine Skizze bei Allison Muri, "Enligthenment Cybernetics: Communication and Control in the Man-Machine," *The Eighteenth Century*, 49 (2008), 141-163, hier insbes. zu Thomas Willis' *Cerebri Anatome* und zu Thomas Hobbes, 142-149.

28) Smith, *Horological Dialogues*, 4.

## 2. Das Uhr-Modell in den Diskursen

Die begriffliche und metaphorische Pointierung des Uhr-Modells sowie seine strategische Funktion innerhalb unterschiedlicher disziplinärer Argumentationen haben nun zwar durchaus diskursübergreifenden, aber deswegen noch lange keinen diskursunabhängigen Charakter. Das Uhr-Modell als akkumulatives metaphorisches und begriffliches Argument findet sich folglich in unterschiedlichen Wissensfeldern der frühen Neuzeit wieder. Bevor ich dazu übergehe, an ausgewählten Autoren und Texten zu versuchen, die hier einleitend an Quellenbeispielen des 15. Jahrhunderts bereits kurz skizzierten heuristischen Indikatoren des Uhr-Modells als *machina machinarum* exemplarisch auszuführen, möchte ich einige der Diskurse, in denen dieses Modell zum Einsatz kam, wenigstens in Umrissen darstellen. Das ist schon deswegen nötig, da bei vielen Autoren, die ich heranziehen werde, eine auffällige Überschneidung unterschiedlicher Diskurse oder Wissensfelder anzutreffen ist.

1) Der heute noch in der Debatte um Kreationismus und *intelligent design* reaktualisierte und daher wohl prominenteste Diskurs ist derjenige der Physiko-Theologie oder Naturtheologie.[29] In der Physiko-Theologie des 17. und 18. Jahrhunderts wurde auf das Uhr-Modell zurückgegriffen, um am Verhältnis von Uhrmacher und Uhrwerk dasjenige von Schöpfer und Schöpfung zu exemplifizieren. Die Welt schien in ihrer planmäßigen Konstruiertheit die Spuren ihres *opifex divinus* genauso an und in sich zu tragen wie das kunstvolle Uhrwerk die Signatur seines technisch und mathematisch hochversierten Konstrukteurs. So hat u. a. der englische Geistliche und mechanistische Naturphilosoph William Derham diesen Vergleich in seiner *Astro-Theology* von 1714 bemüht:

> These and many other Instances, and in a word, that every planet should have as many, and various Motions, and those as regular, and well contrived and ordered, as the World and its Inhabitants have occasion for, what could all this be but the

[29] Vgl. zur Geschichte der Kreationismus Ronald L. Numbers, *The Creationists. From Scientific Creationism to Intelligent Design* (Cambridge, Mass./London, 2006). Populär wurde vor allem die gegen die *New Intelligent Design*-Bewegung argumentierende Schrift Richard Dawkins *The Blind Watchmaker* (New York, 1986).

work of a wise and kind, as well as omnipotent Creator, and Orderer of the World's affairs? A Work which is plain a signal of God, as that of a Clock, or other Machine is that of Man.[30]

Der physikotheologische Diskurs enthielt freilich zwei ganz unterschiedliche Deutungen dessen, was die Welt als vom göttlichen Uhrmacher konstruierte Uhr sein konnte: Zum einen eine unbelebte, rein mechanisch fortlaufend funktionierende Maschine (im Sinne des Deismus vom göttlichen Uhrmacher bei der Schöpfung nur einmal aufgezogen), zum andern ein *horologium animatum* (im Sinne der Lehre von der *creatio continua* nur durch die sich fortlaufend aktualisierende Schöpferkraft Gottes in den Seelenkräften der Natur funktionierend).[31]

---

30) Zitiert nach William Derham, *Astro-Theology Or a Demonstration of the Being and Attributes of God, from a Survey of the Heavens* (London, 1721), 73. Vgl. a. ebd., 124 f.: "When we see divers Pieces of curious Device and Workman-Ship to bear the same marks of Art, to have the same masterly strokes of Painting, Clock-work, Architecture &c. we conclude with great reason such Pieces were made by the same skilful Hand. So when we see the same commodious Spherical Figure to be imparted to the Earth and all the Heavenly Bodies, we have as good reason to conclude them to be Pieces of the same Hand, Contrivances and Works of the same skilful Architect." – Derham verfasste auch ein vielfach neu aufgelegtes Werk zur Horologie, *The Artificial Clock=maker. A Treatise of Watch, and Clock-work: Wherein the Art of Calculating Numbers For most sorts of Movements Is explained to the capacity of the Unlearned. Also the History of Clock-work, both Ancient and Modern* (London, 1696), sowie ein breit rezipiertes naturtheologisches Werk, *Physico-Theology Or, A Demonstration of the Being and Attributes of God, From Hist Works Of Creation […]* (London, 1713), in dem er u. a. ausführlich auf Giovanni Alfonso Borellis *De motu animalium* (Rom, 1680-1681) rekurriert und sich in den naturgeschichtlichen Forschungen seiner Zeit als ausgesprochen versiert erweist. Später wird William Paley in seiner *Natural Theology; Or, Evidences of the Existence and Attributes of the Deity, Collected From the Appearances of Nature* (London, 1811) das Uhr-Argument Derhams noch weiter ausbauen.

31) Für weitere Beispiele (Joachim Rheticus, Johannes Kepler, Robert Boyle, Christian Wolff, Voltaire) siehe Otto Mayr, „Die Uhr als Symbol für Ordnung, Autorität und Determinismus,“ in *Die Welt als Uhr*, 2-6. Ludwig Oechslin, *Die Uhr als Modell des Kosmos und der astronomische Apparat Bernardo Facinis* (Rom, 1985), nennt die gleichen bei Mayr zitierten Beispiele (S. 74 ff.). Oechslin bietet allerdings eine detaillierte Beschreibung und eingehende Analyse der Bedeutung der Farnesianischen Uhr, von Bernardo Facini 1725 in Piacenza für Dorothea Sophia, letzte Herzogin des Hauses Farnese, konstruiert.

2) Das Uhr-Modell fungierte im theologischen Diskurs als metaphorische Vermittlung zwischen Schöpfer und Schöpfung (Ewigkeit und Zeit) und konnte, wie nachher noch am Beispiel von Cusanus nachzuweisen sein wird, Teil und akkumulativer Ausdruck komplexer metaphysisch-theologischer Spekulationen sein. Man konnte von ihm aber auch als sinnbildliche Richtlinie, Wegweiser und als mahnender Weckruf christlicher Devotion und Frömmigkeit Gebrauch machen, wie das in der frühen Neuzeit weitverbreitete Werk des mittelalterlichen Mystikers und Dominikaners Heinrich Seuse, das *Horologium Aeternae Sapientiae* (entstanden um 1323; gedruckt u. a. in Köln, 1496), oder auch das sich am 24-Stundentag orientierende *Horologium devotionis circa vitam Christi* (gedruckt u. a. 1488/9; deutscher Titel: *Zeitglöcklein des Lebens und Leidens Christi*, u. a. Reutlingen, 1494) von Berthold der Bruder, der um 1400 gelebt haben dürfte,[32] belegen. Noch Ende des 17. Jahrhunderts finden sich ähnliche Traktate, wie etwa das *Horologium Christianum* (Oxford, 1689) des englischen Jane Leade-Anhängers Francis Lee.[33]

---

32) Vgl. zu Berthold der Bruder Hans Rupprich, *Vom späten Mittelalter bis zum Barock. Erster Teil: Das ausgehende Mittelalter, Humanismus und Renaissance 1370-1520*, in *Geschichte der deutschen Literatur*, Bd. IV/1, Hg. Helmut de Boor u. Richard Newald (München, 1994), 341; s. a. zur Gattung der Gebets- und Andachtsbücher, unter die die Stundenbücher (*Horaria*) zu zählen sind, ebd., 340-342. Zu den Stundenbüchern im Allgemeinen vgl. u. a. Roger S. Wieck, *Time Sanctified. The Book of Hours in Medieval Art and Life* (New York, 1988).

33) Hier heißt es S. 36-37: "Christus *Horologiarius* Horologium Inspicientem alloquitur; *Christiane* Cur hic stas otiosus, Nonne sunt *duodecim horae* Diei? Me sequere statim: Hoc ipso momento sequere. Fugit momentum: Fugit hora. En *Horologium*! Aspice. Quid moraris? Quid tempus perdis? Hoc *nunc* tuum est. *Nunc* igitur ad me confuge. Nunc sub meo vexillo milita: Ora & labora ut *horam* Mortis beatam aspicias. Crucem tolle, & disce Mori. Sed ut Moriaris disce Vivere/Orare/Credere. Mihi *credas*, qui sum in me *credentibus* Via,. Veritas, & Vita; & te omnibus horis *Credere/Orare/Vivere/Mori* docebo. Sic tibi quoque, si velis, futurus Via, Veritas, & Vita. Ut *moriendo*, sicut & ego *Vivas*, sicut & ego. In Omnem aternitatem Beatus." ("Der uhrentragende Christus spricht, während er auf die Uhr schaut: Christ, warum stehst du hier so müßig, hat denn der Tag nicht zwölf Stunden? Folge mir umgehend: Folge mir genau in diesem Augenblick: Der Augenblick flieht, die Stunde flieht. Sieh nur, die Uhr! Schau sie an. Was zögerst du? Was verlierst du noch Zeit? Dieses Jetzt, es gehört dir. Fliehe also *jetzt* zu mir. Werde *jetzt* Soldat unter meiner Flagge: Bete und arbeite, um die glückselige Stunde des Todes zu erblicken. Nimm das Kreuz auf dich

Für die Verbindung neuer Technologien, der astronomisch-mathematischen *artes* und christlich-religiöser Praxis in der frühen Neuzeit bot sich das Uhr-Modell sicherlich als eines der attraktivsten Versinnbildlichungen an.

So bedankte sich zum Beispiel Melanchthon in einem Brief an den Mathematiker und Mechaniker Georg Hartmann vom 8. August 1542 für eine mit einem Kruzifix versehene Uhr mit den Worten:

> Dort [in den *artes*] zeigen auch die Zahlen, die der Ordnung unseres Verstandes angehören, und die Harmonie der Natur, dass die Gesamtheit der Dinge nicht durch Zufall entstanden ist. Sie zeigen die Gewissheit der logisch-rationalen Kalkulation besonders in der Regierung des Lebens. Daher danke ich Dir für Deine wunderschönen Werke und Geschenke. Die Übereinstimmung des Magneten mit dem arktischen Pol, die ein großes Wunder ist, wird mich, wenn ich in Dein Werk schaue, oft an die Übereinstimmung unserer Seelen mit der himmlischen Heimat erinnern. Auch das der Uhr angehängte Kreuz wird mich daran erinnern, dass irgendwann das Ende dieser elenden Zeiten kommen wird und dieser unser Magnet, das ist: unsere Seelen, in jene seine himmlische Heimat zurückkehren wird.[34]

---

und lerne zu sterben. Aber um zu sterben, lerne Leben/Beten/Glauben. Du sollst an mich glauben, der ich den an mich Glaubenden Weg, Wahrheit und Leben bin; und ich werde dich zu allen Stunden Glauben/Beten/Leben/Sterben lehren.Und so werde ich auch dir, wenn du willst, Weg, Wahrheit und Leben sein. Damit du, im Sterben wie ich, lebst wie ich. Glückselig bis in alle Ewigkeit.").

34) Philipp Melanchthon, "Epistolarum Lib. IX," in *Corpus Reformatorum, Philippi Melanthonis Opera quae supersunt omnia,* Hg. Carl Gottlieb Bretschneider, Bd. 4 (Halle, 1837), 853-854: "Ibi numeri et ordinis intellectus, et naturae harmonia ostendunt, non casu extitisse hanc rerum universitatem, et monstrant certitudinem ratiocinationis in gubernatione vitae praecipua. Itaque pro tuis pulcherrimis operibus et donis tibi gratias ago. Et magnetis consensus cum polo Arctico, qui magnun miraculum est, saepe me intuentem tuum opus monebit de consensu nostrarum animarum cum coelesti Patria. Et crux addita Horologio monebit harum miseriarum finem aliquando fore, et hunc nostrum magnetem, hoc est, animas nostras in illam suam Patriam coelestem redituras esse." – Geschenke dieser Art waren keine Seltenheit. Auch Luther bedankte sich in einem Brief vom 22. April 1527 bei einem Uhrmacher namens Friedrich Pistorius für eine metallene Uhr, die dieser ihm zum Geschenk gemacht hatte. Freilich müsse er erst „Schüler unserer Mathematiker" werden, „bis ich alle Regeln und Formeln dieser einzigartigen Uhr verstehe"; Martin Luther, *Gesammelte Werke*, Hg. Kurt Aland, Bd. 10, 178.

Die mit einem Kreuz versehene Uhr,[35] von Melanchthon fast nur beiläufig erwähnt, wird in seinem Brief an Georg Hartmann zum Sinnbild des heilsgeschichtlich zirkulären Weges der menschlichen Seele, zum Mahnmal der Erlösung aus dem Elend der irdischen Zeit in die himmlische Ewigkeit durch die *magnetische* Gnade Gottes. Zugleich ist die Uhr ein die Innovationskraft der mathematischen *artes* eindrucksvoll belegendes Beispiel, Signum der dem menschlichen Geist einwohnenden *numeri*, folglich der *zählenden* und *messenden* Ordnungskraft des menschlichen Verstandes, die in einem spiegelbildlich-analogen Verhältnis zur göttlichen, nach Maß, Zahl und Gewicht eingerichteten Ordnung der Natur steht. Nur vor diesem Hintergrund ist auch die logisch-rationale Regulation und Beherrschung des Lebens durch Kalkulation möglich. Als ein sozialpolitisch relevantes Resultat einer solchen rationalen Kalkulation dürfte Melanchthon die das alltägliche Leben – an den Schulen, Universitäten etc. – rhythmisierende Uhr angesehen haben, für deren technische Weiterentwicklung mathematisch versierte Mechaniker wie der Adressat des Melanchthon-Briefes seinerzeit berühmt waren.

3) Eng verbunden mit dem naturtheologischen Diskurs ist der kosmologische und astronomische, schließlich mit und seit Descartes nachhaltig vom Mechanismus geprägte Diskurs, in dem Themen wie die qualitative und quantitative Zahlstruktur des Kosmos, das naturmagisch oder naturgesetzlich verfahrende göttliche Weltregiment, Kosmogonie (Entstehung der Planeten etc.) sowie die Welt als nach festen Regeln oder Gesetzen funktionierende *machina* verhandelt wurden. Auch hier war die Uhr-Metapher geeignet, die Qualitäten der Weltmaschine (*machina mundi*) sinnbildlich vor Augen zu führen. Ihre Verwendung verweist zugleich auf die zunehmende Wichtigkeit mechanischer Künste in der frühen Neuzeit.[36]

4) Sicherlich begünstigt durch das in der frühen Neuzeit flottierende kosmologisch-anthropologische Mikrokosmos-Makrokosmos-Schema

---

35) Vgl. mögliche vergleichbare Uhren in *Die Welt als Uhr*, 183 (Katalognr. 20: Tischuhr um 1600 mit aufgesetztem Kreuz), 242 (Katalognr. 68: Figurenuhr um 1640 mit Darstellung der Kreuzigung), 243 (Katalognr. 69: Figurenuhr um 1580/90 mit Darstellung des Kalvarienbergs).

36) Vgl. dazu auch Domenico Bertoloni Meli, *Thinking with Objects. The Transformation of Mechanics in the Seventeenth Century* (Baltimore, 2006).

ließ sich an den kosmologisch-astronomischen Diskurs unmittelbar der Körper-Maschine-Automat-Diskurs anschließen, in dem Automaten, Maschinen und Horologia u. a. zum technomorphen Modell für den organischen Körper avancierten.[37] Dieser Diskurs war freilich kaum von der Frage zu trennen, wie das Verhältnis und die Funktionsweisen von Seele, Geist und Körper zu bestimmen sind, so dass er drei eng miteinander verflochtene Wissensfelder, nämlich Physik, Erkenntnistheorie und Psychologie, berührte.[38]

5) Ein weiterer Diskurs drehte sich um die *Antiqui* und *Moderni*, um den breit diskutierten Stellenwert des Neuen, der *novitas*, des

---

37) Vgl. dazu den Überblick bei Renato G. Mazzolini, „Mechanische Körpermodelle im 16. und 17. Jahrhundert," in *Technomorphe Organismuskonzepte. Modellübertragungen zwischen Biologie und Technik*, Hg. Wolfgang Maier u. Thomas Zoglauer (Stuttgart-Bad Cannstatt, 1994), 113-133. Der These Mazzolinis (ebd., 126), die Entwicklung von Automaten und Uhrwerken hätten dazu geführt, dass architektonische Metaphern durch komplexere Maschinenmodelle abgelöst worden seien, um die Funktionsweise organischer Körper zu veranschaulichen, kann ich so nicht zustimmen. Beide Metaphern – die Uhr-Metapher und die Architektur-Metapher – kommen m. E. in der frühen Neuzeit weiterhin parallel vor. Die eine musste die andere nicht unbedingt ablösen, da beide unterschiedlichen Logiken folgen. Meine These: Die Anatomie bedient sich eher der statischen Logik der Architektur, während sich die Physiologie eher der dynamischen Logik der Horologie bedient. Im naturtheologischen Diskurs kommen beide Metaphern gleichwertig zum Einsatz, da sie hier dem meist nicht weiter differenzierten Oberbegriff und Argument der handwerklichen Konstruiertheit (der Welt, des Gebäudes und der Maschine) subsumiert werden können.

38) In Diskussionen zur künstlichen Intelligenz und in modernen neurophysiologisch-kybernetisch bestimmten kognitionswissenschaftlichen und psychologischen Positionen, die für die heutige Frage nach der Freiheit des Menschen und einer Neubegründung der Ethik zentral sind – dafür stehen beispielhaft die Debatten in der sogenannten Neurophilosophie –, hat sich dieser Diskurs ganz im Sinne Foucaults reaktualisiert. Vgl. z. B. Henrik Walter, *Neurophilosophie der Willensfreiheit. Von libertarischen Illusionen zum Konzept natürlicher Autonomie* (Paderborn, 1998) und neuestens Paul Churchland, *Neurophilosophy at Work* (Cambridge, 2007). Vgl. außerdem die interessante Studie von Allison Muri, *The Enlightenment Cyborg. A History of Communications and Control in the Human Machine, 1660-1830* (Toronto, 2007); Tim Crane, *The Mechanical Mind: A Philosophical Introduction to Minds, Machines, and Mental Representation* (London, 1995); sowie die Sammelbände *Mechanical Bodies, Computational Minds: Artificial Intelligence from Automata to Cyborgs*, Hg. Stefano Franchi u. Güven Güzeldere (Cambridge/Mass., 2005) und *The Mechanical Mind in History*, Hg. Philip Husbands (Cambridge/Mass., 2008).

technologischen und ökonomischen Fortschritts, um die mit diesem verbundene politische Macht und um das Verhältnis des Neuen und Innovativen zu Traditionen, die als altehrwürdig empfunden wurden, sowie zu gewohnten und bewährten praktischen Methoden in einzelnen Wissensfeldern, z. B. in der Medizin. Innerhalb der Diskussion um Invention und Innovation war die technische Weiterentwicklung der Uhr zwar nur ein Argument oder Beispiel unter vielen, aber doch neben dem Buchdruck aufgrund ihrer alltäglichen Präsenz ein besonders exponierfähiges, wie an den bereits oben zitierten Autoren – Tortelli, Poliziano, Vergil – deutlich wird. Buchdruck und Horologium bestachen vor allem mit ihrer immer wieder gerne betonten Nützlichkeit für das Gemeinwohl.[39] Dies war Anlass genug, sie als schlagkräftige Argumente zu verwenden, wenn es darum ging, ein als *neu* apostrophiertes Wissen politisch zu propagieren oder gegen Anfeindungen in Schutz zu nehmen. So begründete und rechtfertigte der Paracelsist Adam von Bodenstein den gemeinnützigen Aspekt der paracelsischen *medicina nova* mit dem Hinweis auf den großen Nutzen der neu invenierten „horologia“, „ars typographica“ (Buchdruck) und des „compassus“.[40] Die neu invenierten technologischen Errungenschaften signalisierten den gottgewollten Fortschritt, der die 'Moderne' gegenüber der heidnischen Antike auszeichnete. Dies implizierte jedoch keineswegs einen radikalen Bruch mit der Tradition: Das Verhältnis zur Tradition kann nämlich sowohl progressiv – durchaus im Sinne der relativen Höherbewertung der 'Moderne' gegenüber der Antike – als auch typologisch bzw. genealogisch gekennzeichnet sein.

In seiner Schrift *Warhafftige Außlegung des Astronomischen Uhrwercks zu Straßburg* (Straßburg, 1578) rekapituliert Cunradus Dasypodius die seinerzeit breit diskutierte Frage wie folgt:

---

39) Vgl. Dohrn-van Rossum, *Geschichte der Uhr*, 181 (mit Quellenbelegen): „Die öffentliche Uhr war die 'urbanistische Investition par excellence', und da hier praktischer und symbolischer Nutzen zusammenfielen, brauchte in der Regel nicht gespart zu werden.“

40) *Corpus Paracelsisticum. Dokumente frühneuzeitlicher Naturphilosophie in Deutschland*, Bd. I, *Der Frühparacelsismus. Erster Teil*, Hg. Wilhelm Kühlmann u. Joachim Telle (Tübingen, 2001), 114.

> Es ist ein frag vnder den gelehrten / die ob sie wol nicht sehr wichtig scheint sein / so hat sie doch ein gůt vnd lustig bedencke[n]/ namlich ob gelehrtere vnd verstendigere, auch künstlichere leüt / vor alten zeiten / bey den Juden / Griechen / Römeren / vnd anderen völckeren / gewesen seyen / oder aber jetz zu vnseren zeite[n] / das diejenigen so in guten freyen künsten sich jetziger zeit üben / vnd gelehrte künstliche leut seindt vnd geachtet werde[n], höheres verstandts vnnd geschicklichkeit seye dann die alten.[41]

Dasypodius sieht die ‚Moderne' der Antike verpflichtet, so dass erstere nichts als eine Ausgestaltung bruchstückhaft tradierter Vorgaben der Antike ist. Nach Dasypodius würde man den antiken Meistern, wie Heron von Alexandria oder Archimedes, Unrecht tun, wenn man sie allein nach den unvollständig überlieferten Quellen beurteilen würde. Ohne Zweifel hat nämlich auch hier die göttliche Vorsehung die Entwicklung der *artes* gefördert, so dass „[d]erhalben darauß endtlich zu schliessen ist / dz kein underscheid ist welche gelehrter oder künstlicher gewesen seyen / die alte[n] oder aber die so zu unser zeit leben […].“[42]

Johann Geiger führt in seinem 1621 veröffentlichten *Horologium politicium* die Uhrmacher auf den alttestamentarischen Sohn Kains, Tubalcain, zurück und begründet damit die 'Moderne' genealogisch in der unmittelbaren Nachkommenschaft Adams:

> Vnter solchen Nachkömlingen TubalCains / seyn heutigs Tags nicht die wenigste vnnd geringste die Uhrmacher / so durch eingebung deß heiligen Geistes / ihren sinnreichen Verstand vnnd Kunst / künstliche Uhrwerck machen vnd außfertigen […], deren grosser hoher Nutz vnd Gebrauch im menschlichen Leben nicht gnugsam außzusprechen ist / vnd derer man im menschlichen Leben / Handel vnnd Wandel / so wenig entrahten vnd entperen mag vnd kan […].[43]

Es ließen sich durchaus noch weitere Wissensfelder nennen, in denen die Uhr-Metapher appliziert wurde. Zu nennen wären etwa der

---

41) Dasypodius, *Warhafftige Außlegung*, Vorrede, Aij^r^.

42) Ebenda, Aiv^r^-Aiv^v^.

43) Johann Geiger, *Horologium politicum. Das ist: Einfältiger Discurs und kurtze Betrachtung / deß politischen / materialischen / selbst gehenden vnnd schlagenden Uhrwercks / desselben Nutz und Gebrauch / zu sampt andern Automatibus vnd selbst bewegenden Wercken / so auß solchem entstehen / vnd demselben verwandt vnd angehörig […]* (Nürnberg, 1621), Dedication-Schrifft, Aij^v^.

politische und der moralisch-pädagogische Diskurs.[44] Festzuhalten aber bleibt, dass die Uhr deswegen eine so große Wirkung entfalten und eine solch akkumulative argumentative Funktion in den einzelnen Diskursen haben konnte, weil sie ein innovatives, technisches, politisch-ökonomisches, regulatives und prestigeträchtiges Instrument der alltäglichen Lebenswelt der frühen Neuzeit war. Dadurch erst ist es möglich gewesen, dass sie als Begriff, Metapher und Exempel unterschiedliche Diskurse miteinander in Beziehung zu setzen und zu vernetzen vermochte.

Im Folgenden möchte ich nun an einigen Autoren – Nicolaus Cusanus, Marsilio Ficino, René Descartes, Gottfried Wilhelm Leibniz und Christian Wolff – die diskursübergreifende Verwendung des Uhr-Begriffs und der Uhr-Metapher zwischen 1450 und 1750 in aller Kürze veranschaulichen.

## 3. Das horologische Vergleichsmodell in Nikolaus von Kues' *De Visione Dei* (1453)

Soweit ich sehe, verwendet Nicolaus Cusanus den Begriff der Uhr nur singulär, nämlich an einer zentralen Stelle seiner von Kurt Flasch als „der poetischste Text" des Cusaners bezeichneten Schrift *De visio Dei* von 1453,[45] um das Verhältnis von Ewigkeit und Sukzession, von Einfaltung und Entfaltung (*complicatio* und *explicatio*), von Begriff (*conceptus*) und Realisierung des Begriffs zu veranschaulichen.[46] Der Begriff der Uhr stellt dabei eine besondere Pointe der bis zu seinem Auftauchen im Text fromm explizierten Argumentation dar. Dabei ermöglicht das

---

44) Als einschlägiges Beispiel für den moralischen Diskurs, der sich hier mit dem politischen überschneidet, nenne ich: Antonio de Guevara, *Reloj de Príncipes* (Vallodolid, 1529); die lateinische Übersetzung *Horologium Principum* erlebte zahlreiche Auflagen in der frühen Neuzeit und in der Aufklärung und wurde ins Englische, Französische und Deutsche übertragen. – Eine technische Abhandlung über die Uhr, die deren sozialpolitischen Nutzen herausstellt, bietet der schon genannte Johann Geiger mit seinem *Horologium politicum*.

45) Kurt Flasch, *Nikolaus von Kues. Geschichte einer Entwicklung* (Frankfurt/Main, 2008; 1. Auflage Frankfurt/Main, 1998), 385.

46) Vgl. zu den Grundtheoremen des Cusaners, zu denen das *complicatio-explicatio*-Schema gehört, Thomas Leinkauf, *Nicolaus Cusanus. Eine Einführung* (Münster, 2006), 89 ff.; zum *complicatio-explicatio*-Schema vgl. ebd., 102-110.

in sich hochkomplexe, konsistente metaphysisch-theologische System des Cusaners einen für die Bedeutung der Uhr aufschlussreichen Rückbezug der späteren Schrift *De ludo globi* auf das horologische Gleichnis in *De visione Dei*.

Ausgangspunkt von *De visione Dei* ist die Analogie, die Cusanus zwischen der Allgegenwart des göttlichen Sehens und der Wirkung zeitgenössischer Porträtmalerei zieht. Das Phänomen, dass die *picta imago* eines Gesichts den Eindruck zu erwecken vermag, das gemalte Angesicht fokussiere den Betrachter unabhängig von dessen Standpunkt selbst dann, wenn dieser sich bewege, interpretiert Cusanus, wie später im Text das Horologium, als Vergleichsmodell für die transzendente Präsenz oder Immanenz der Ewigkeit Gottes in der Zeit. Obwohl sich das göttliche Sehen genausowenig bewegt wie das porträtierte Angesicht, ist es doch – über die Gegensätze hinaus und diese zugleich grundierend – der Bewegung oder Ruhe des Stand- oder Perspektivpunkts aller Betrachter jederzeit präsent. Während dies für das Verhältnis eines bestimmten Betrachters zu einem bestimmten Einzelporträt in nur beschränkter, nämlich zeitlich und räumlich singulär realisierter Weise zutrifft – ein Verhältnis, das nur *eine* entfaltete bzw. realisierte Möglichkeit der im göttlichen Geist enthaltenen Ideen des absoluten Angesicht-Sein- und Sehen-Könnens von unendlich vielen entfaltbaren Possibilitäten vorstellt –, gilt für das Verhältnis Gottes zur Schöpfung, dass Gott die urbildliche *complicatio absoluta* ist, die Einfaltung alles Möglichen schlechthin, aus der heraus Gott sich als Prinzip, Ursache und Kraft in die Schöpfung überall und zu jeder Zeit abbildet, expliziert und offenbart, ohne dabei seine Transzendenz einzubüßen.[47]

Noch bevor Cusanus auf das Vergleichsmodell des Horologium rekurriert, exemplifiziert er die Relation zwischen Kraft, Eingefaltetsein und Entfaltetsein am Beispiel des Nussbaums (*arbor nucus*). Zentral für die Bestimmung der kosmologisch und theologisch fundamentalen Relation von *complicatio* und *explicatio* ist die Verbindung von sinnlicher

47) Nicolaus Cusanus, *De visione Dei*, in *Philosophisch-Theologische Schriften*, Hg. Leo Gabriel, übersetzt v. Dietlind und Wilhelm Dupré, Bd. III (Wien, 1989), 104: Gott als "absolutum esse omnium", 116: "Nam cum tu sis vis illa Domine seu principium ex quo omnia [...]."; 120: Gott als Alles in Allem: "[...] qui tu es omnia in omnibus"; 124: das göttliche Sehen als universale Kausalität: "[...] quia videre tuum est causare. Omnia vides, qui omnia causas"; 126: Gott als "causa omnium visibilium".

und mentaler Erkenntnis: Mit dem sinnlichen Auge (*oculus sensibilis*) werden die nur *hier und jetzt* wahrnehmbaren spezifischen Qualitäten eines *bestimmten* Nussbaums erfasst, mit dem geistigen, am göttlichen Intellekt partizipierenden Auge (*oculus mentis*) wird hingegen die Tatsache erkannt, dass dieser bestimmte Nussbaum nicht immer so gewesen ist und auch im Samen keineswegs so enthalten war, wie er *hier und jetzt* vor dem sinnlichen Auge in Erscheinung tritt. Im Samen waren nicht nur der dann konkret realisierte Nussbaum, sondern auch mit den im Samen bereits angelegten neuen Nusssamen (wie wiederum die in jedem einzelnen dieser neuen Nusssamen enthaltenen weiteren Nusssamen usw.) potentiell unendlich viele Nussbäume eingefaltet, so dass Cusanus aus der unendlichen seminalen Verschachtelung schließt, die seminale Kraft eines einzigen Nussbaumsamens an sich sei „zu keiner Zeit in und mit der Bewegung des Himmels in seiner gesamten Fülle entfaltbar.“[48] Und doch ist die in der Zeit unausschöpfbare Seminalkraft des Nussbaumsamens insofern eingeschränkt, als sie auf eine bestimmte Baumart (*in hac specie nucum*) beschränkt (*contracta*) ist. Was aber letztlich auf etwas Bestimmtes hin *kontrahiert* wird – wie der unbewegliche Blick des porträtierten Angesichts, der auf jeden einzelnen Betrachter konzentriert resp. *kontrahiert* zu sein scheint –, ist das absolute göttliche Prinzip, die omnipotente göttliche Kraft selbst, „welche allen seminalen sowie nicht-seminalen Kräften ihr Sein gibt“[49] und von Cusanus als absolute, d. h. von jeder besonderen *virtus* losgelöste urbildliche, jegliches Kreatürliche konzipierende Kraft der Kräfte ausgewiesen wird (*virtus virtutum absoluta*).

Die Offenbarung des Nicht-Zeitlichen in der Zeit, des ewig Gleichzeitigen und in sich Verharrenden in der Bewegung und im Leben wird zwar in den folgenden Kapiteln immer wieder angerissen, aber erst im elften Kapitel zum zentralen Thema der Schrift erhoben. Darin beruft sich Cusanus auf eine *similitudo grata*, auf ein „willkommenes Gleichnis“, das dazu geeignet sei, noch tiefer in das Geheimnis des Verhältnisses von Welt und Schöpfer, von sukzessiver Entfaltung des göttlichen Wortes in der Schöpfung und des ewigen Eingefaltetsein der Ideen oder

---

48) Cusanus, *De visione*, 118: “[…] non est ullo umquam tempore motu caeli ad plenum explicabilis […].”

49) Ebenda: “Quae [die göttliche *virtus stupendissima*] est principium dans esse omni virtuti seminali et non seminali.”

Begriffe im Geist Gottes einzudringen. Für dieses von Cusanus besonders hervorgehobene Gleichnis steht nun der Begriff und die Metapher der Uhr:

> Du hast mir ein willkommenes Gleichnis eingehaucht [*inspirasti*], das sich um die Einheit Deines geistigen Wortes bzw. Begriffes und dessen Mannigfaltigkeit in all dem, was sukzessiv in Erscheinung tritt, dreht. Denn der einfache Begriff der vollkommensten Uhr führt mich dahin, dass ich noch wissender zum Sehen Deines Begriffes und Wortes hingerissen werde.[50]

Cusanus entwickelt die Uhr-Metapher daraufhin in zwei Schritten: (1) Die Uhr als Begriff der zeitlichen Sukzession schlechthin; (2) die Uhr als Begriff der Ewigkeit selbst. Bereits an diesen beiden Definitionen des Uhr-Begriffs zeigt sich, dass Cusanus am mechanischen Vergleichsmodell des Horologium die göttliche *coincidentia oppositorum* – hier die Koinzidenz der Gegensätze Ewigkeit-Zeit, permanente Gleichzeitigkeit-Sukzession, Ruhe-Bewegung – sowie das kausale Verhältnis von göttlicher Ewigkeit und weltlicher Zeit, von Geist, Seele und Körper, von Inbegriff und Objekt veranschaulichen möchte.

Zunächst definiert Cusanus die Uhr wie folgt: „Denn der einfache Begriff der Uhr faltet alle zeitliche Sukzession in sich ein.“[51] Die Einfachheit des Begriffs der Uhr meint die Unteilbarkeit seiner divinen Konzipierung, die nicht nach zeitlich-räumlichen Kriterien in ein Früher oder Später, Vor oder Hinter aufgeteilt werden kann. Der mentale Begriff oder der gedankliche Plan der Uhr, den sich der göttliche Uhrmacher von der Uhr macht, ist, obwohl er die Idee der Zeitfolge und -angabe beinhaltet, nicht selbst der konkreten Zeitfolge der hier von Cusanus anvisierten stundenschlagenden Uhr im Sinne eines Früher oder Später unterworfen, wie das etwa in der Diskursivität des menschlichen Verstandes der Fall ist, der am Begriff im göttlichen Geist partizipiert, indem er diesen sukzessive in seine Bestimmungen oder, um

---

50) Ebenda, 138: “Inspirasti similitudinem mihi gratam circa unitatem verbi mentalis seu conceptus tui et varietatem eiusdem in successive apparentibus. Nam simplex conceptus horologii perfectissimi me ducit, ut sapidius rapiar ad visionem conceptus et verbi tui.”

51) Ebenda, 138: “Conceptus enim simplex horologii complicat omnem successionem temporalem.”

mit Cusanus zu sprechen, in seine Kontraktionsmomente zerlegt. Vielmehr ist der absolut einfache Begriff der Uhr, neben seinem Status als Bestand der Weisheit Gottes, zugleich und wesentlich göttlicher Wille und Befehl. Als solcher stellt er die gleichsam metaphysische *machina machinarum* vor und ist Bedingung jeder konkret – und sei es im Verstand eines Uhrmachers – konstruierten Uhr, die nur deswegen ist, weil sie am (Soll)Plan des Meisters (*conceptus magistri*) teilhat. Der Plan selbst birgt eine unhintergehbare Kraft, die sich in der Entstehung und Konstruktion der konkreten Gegenstände als ein Sein-Sollen nach Vorgabe des Konzepts äußert. Der Begriff der Uhr meint folglich nicht, dass in ihm selbst schon, wie Cusanus schreibt, die sechste Stunde (*sonus sextae horae*) *tatsächlich* früher angezeigt wird als die siebte Stunde, sondern dass alle Stunden im Begriff der sich aus der Einheit entfaltenden Zahlenfolge absolut und noch unausgefaltet, *virtualiter*, enthalten sind – darunter auch, verstanden als ein auf die Uhr-Idee beschränktes Werden-Können, die Möglichkeit, dass die sechste Stunde vor der siebten Stunde ertönt. Somit ist der Begriff der Uhr die von jeder Konkretion losgelöste (absolute) Ur-Sache für die unterschiedlichen horologischen Realisierungsformen in Zeit und Raum genauso wie die Ein(s)heit die *causa numerorum* ist, die Ursache der Zahlen, denen sie, obwohl und weil nicht selbst Zahl und Folge, jederzeit gegenwärtig ist. Die Uhr repräsentiert demnach die Entfaltung der Zahlen aus dem Geist der göttlichen Einheit sowie die mechanische Transformation der Zahlenfolge – und damit der Einheit – in Bewegung (*motus*) nach dem befehlenden Willen des Konstrukteurs, der den allgemeinen Bauplan entworfen hat.

In einem zweiten Schritt, der das Vorhergehende zusammenfasst und beschließt, spezifiziert Cusanus die Uhr-Metapher, indem er sie zum Sinnbild der Ewigkeit erklärt:

> Also soll der Begriff der Uhr gewissermaßen die Ewigkeit selbst sein, die Bewegung in der Uhr ist dann die Sukzession. Demnach faltet die Ewigkeit die Sukzession zugleich ein und aus. Denn der Begriff der Uhr, der die Ewigkeit ist, faltet zugleich alles ein und aus.[52]

---

52) Ebenda, 140: "Sit igitur conceptus horologii quasi ipsa aeternitas, tunc motus in horologio est successio. Complicat igitur aeternitas successionem et explicat. Nam conceptus horologii, qui est aeternitas, complicat pariter et explicat omnia."

Wie bereits oben ausgeführt, impliziert der Begriff der Uhr das virtuelle Werden-Können der zeitlichen Sukzession, ohne selbst von dieser zu dependieren. Die an dieser Stelle von Cusanus im Kontext des horologischen Vergleichsmodells neu herangezogenen Begriffe der Ewigkeit und der Bewegung ermöglichen es, die ständig tätige Dynamik von *complicatio* und *explicatio* näher zu umschreiben. Das zeitliche Prinzip der Sukzession drückt sich in der kontinuierlichen, zirkulären Bewegung des Räderwerks im Inneren der Uhr aus, die in der äußeren Bewegung des Zeigers von Zählstunde zu Zählstunde gespiegelt wird. Die Verschränkung von Innen und Außen, von innerem Mechanismus und äußerer Signalwirkung ergibt zusammen die für sich als Ganzes ruhende Uhr als *Bild* der Ewigkeit. Diese selbst ist die produktive permanente Gleichzeitigkeit zeitloser Begriffe im Geist Gottes mit allem, was davon raum-zeitlich realisiert bzw. kontrahiert war, ist und sein wird. Sie ist produktiv-entfaltend, insofern sie zugleich die treibende Kraft des Zusammenziehens eines Allgemeinen zu etwas konkret Bestimmten ist; sie ist komplikativ, insofern sie im konkret Bestimmten dessen ursächlich Allgemeines signalisiert und endlich auf sich selbst als jenes Prinzip zurückführt, das absoluter Träger des Allgemeinen ist: Gott als *aeternitas* selbst.

Mit der Uhr als Bild der Ewigkeit und der inneren sukzessiven Bewegung des Uhrwerkmechanismus korrespondieren bei Cusanus nun noch der Kreis, die Kreisbewegung und die Seele. Der Kreis ist für Cusanus die vollkommene geometrische Figur *par excellence*. Im Kreis gibt es, wie es in der um 1462/1463 entstandenen Schrift *Dialogus de ludo globi* heißt, keinen Anfang und kein Ende, „da in ihm kein Punkt vorhanden ist, der eher Anfang als Ende wäre [...]." Daher sieht Cusanus im Zirkel, wie vorher im Horologium, eine „imago aeternitatis", ein Bild der Ewigkeit, das an der zeitlosen Gleichzeitigkeit der Anfang und Ende, Vorher und Nachher einfaltenden und aufhebenden Ewigkeit Gottes partizipiert.[53] An der Verknüpfung von Kreis und Seele offenbart sich, wie aus dem Folgenden ersichtlich wird, Leben und Lebendigkeit. Die Seele ist bei Cusanus nämlich Prinzip der „sich selbst

---

[53] Cusanus, *Dialogus de ludo globi liber primus*, in *Philosophisch-Theologische Schriften*, Bd. III, 234-236: "In circulo enim, ubi non est principium nec finis, cum nullus punctus in eo sit, qui potius sit principium quam finis, video imaginem aeternitatis." Zu dieser Schrift vgl. Flasch, *Nikolaus von Kues*, 576 ff.

bewegenden Bewegung", die zugleich eine „Bewegung des Verlebendigens" (*motus vivificandi*) ist.[54] Sie ist das den passiven Körper belebende Regulativ, das, ohne an den Körper gebunden zu sein, in allen seinen Teilen immer und gleichzeitig gegenwärtig ist. Als Weltseele verleiht sie den himmlischen Sphären deren kreisförmige Bewegung, die ihrerseits Grundlage der die Bewegung messenden Zeit ist.[55] Die Kreisbewegung, gleichsam die Resultante aus Kreisförmigkeit und Seele, ist somit bei Cusanus mustergültiges Vorbild des Lebendigen schlechthin. Leben ist das in sich bewegte Geschlossene und Runde, das durch menschliche Kunstfertigkeit und Erfindungsgabe nachgeahmt werden kann. Kunst und Technik nehmen Maß am Leben, indem sie dieses simulieren. Zwischen Welt, Kreis, Seele, Leben, Bewegung und Zeit besteht demnach ein verschränktes Kausalverhältnis. Die Seele ist als Prinzip der Bewegung notwendig zugleich Prinzip der die Bewegung messenden Zeit. Das optimale Zeitmaß ergibt sich dann folgerichtig aus der optimalen, d. h. aus der in sich geschlossenen, zirkulären Bewegung. Diese manifestiert sich in der Kreisläufigkeit der animierten himmlischen Weltsphären, der Planeten, deren *revolutiones* sich schließlich in unterschiedliche, instrumentale Zeiteinheiten aufteilen lässt: "So sind Jahr, Monat und Stunden Instrumente des vom Menschen geschaffenen Zeitmaßes. Somit ist die Zeit, weil sie Maß der Bewegung ist, Instrument der messenden Seele."[56]

Daraus ergeben sich, bezieht man die Ausführungen des Cusaners in *De ludo globi* auf die Uhr-Metapher in *De visione Dei* zurück, folgende Bestimmungen des horologischen Vergleichsmodells: Dieses repräsentiert ein in sich geschlossenes, belebtes, da bewegtes, d. h. beseeltes System, das 1. den Seelen-Mechanismus bzw. das Leben der

---

54) Ebenda, 242: "Sed creatus est in te motus se ipsum movens secundum Platonicos, qui est anima rationalis movens se et cuncta sua." Ebd.: "[...] motus vivificandi animal [...]."

55) Vgl. ebenda, 239: "[...] tempore, quod est mensura motus."; 261: "Vis enim illa [= anima] mundi, quae se ipsam et omnia movet, [...] est perpetua, quia motus rotundus et circularis omnem in se habens motum [...]."

56) Cusanus, *Dialogus de ludo globi liber secundus*, in *Philosophisch-Theologische Schriften*, Bd. III, 325: "Sic annus, mensis, horae sunt instrumenta mensurae temporis per hominem creatae. Sic tempus, cum sit mensura motus, mensurantis animae est instrumentum."

Welt simuliert, indem es die Zeit, d. h. die Bewegung der Welt, mittels der Uhrwerkbewegung nachbildet und zugleich misst; 2. dadurch zum Signum der messenden, zählenden, selbst- sowie den physischen und sozialen Körper regulierenden Tätigkeit der rationalen Seele des Menschen wird, die sich u. a. in der Invention neuer Instrumente äußert;[57] und 3. in seiner (naturtheologischen und frömmigkeitspragmatischen) Funktion als technisches Gleichnis für die ewige Gleichzeitigkeit der göttlichen Allgegenwart in der Welt an die *causa prima* aller Dinge erinnert und zu dieser 'glockenschlagend' zurückruft.

Besonders wichtig und signifikant ist die von Cusanus hervorgehobene Analogie der Uhr mit der *anima* als Prinzip der Bewegung und des Lebens sowie mit Gott, der als *virtus virtutum* alles Leben jederzeit und überall mit der Kraft, in Raum und Zeit existieren zu können, ausstattet. Das Horologium steht bei Nikolaus von Kues demnach gerade nicht, wie es später in der mechanistischen Naturphilosophie der Fall sein wird, für das intrinsische, autonome Funktionieren des Körpers selbst. Vielmehr versinnbildlicht es jene immaterielle Kraft und jenes unkörperliche Prinzip, die das Funktionieren des Körpers allererst ermöglichen. Es verweist somit auf das psychologische Modell des Körpers, in dem letzterer keineswegs autonom ist, sondern von der ihrerseits vom Körper unabhängigen Seele und deren *virtus* dependiert. Aber auch die Autonomie der Seele ist nur eine relative Autonomie, die der absoluten Autonomie Gottes subordiniert ist.

Gott ist, denkt man das Uhrengleichnis bei Cusanus konsequent weiter, dem Mittelpunkt des Kreises vergleichbar, um den sich der an ihm aufgehängte, vom Uhrwerk angetriebene Stundenzeiger bewegt.

Der quasi autonome horologische Automat, so lässt sich abschließend zusammenfassen, erscheint in *De visione Dei* als repräsentatives Sinnbild der Seele und Gottes, nicht aber – und das ist hier bezeichnend – des Körpers, dem innerhalb des Gleichnisses allenfalls noch das Uhrgehäuse (die Statik der Uhr) entsprechen mag.

---

57) Ebenda, 325: "Creat anima sua inventione nova instrumenta ut discernat et noscat, ut Ptolemäus astrolabium […]."

## 4. Horologium und *Temperatio* bei Marsilio Ficino

> Wir werden jenes Band aber gerne Seele, ja vielmehr Kraft der Seele nennen, insofern die Kraft der Seele [*virtus animae*] im Körper fast genau dem entspricht, was die Kraft der Gerinnung in der Milch ist, so dass die Milch gerinnt, oder was die Kraft des Magneten im Eisen ist, so dass dieses angezogen und hochgehoben wird, oder was die Temperierung [*temperatio*] durch Kunst in der Uhr ist, so dass diese in wunderbarer Ordnung gelenkt wird, bewegt wird und ertönt. Die Seele aber ist nicht selbst die Kraft oder die Temperierung[58] der Säfte, sondern eine bestimmte substanzielle, einfache und unteilbare Form, die durch ihre natürliche Kraft und Kunst die Geister [*spiritus*], die Säfte [*humores*] und die Glieder temperiert.[59]

Dieses Zitat aus dem sechsten Buch des 1482 in Florenz erschienenen Hauptwerks Marsilio Ficinos, *Theologia Platonica de immortalitate animorum*, ist die einzige Stelle überhaupt, in der Ficino das Horologium als Vergleichsmodell zur Beschreibung des Verhältnisses von Körper und Seele bemüht. Umso beachtlicher ist es, dass es nicht das der Natur entnommene Beispiel der Milch oder des Eisens, sondern das horologische *artificium* ist, das Ficino als Folie dient, um die *virtus animae* näher zu bestimmen: Es ist das in der Uhr wirksame temperierend-regulierende Moment, das von Ficino auf die den Körper organisierende Kraft der Seele übertragen wird. Diese wechselseitige Transferbeziehung wird rhetorisch durch die Seele und Uhr analogisierenden Begriffe der *ars* und der *temperatio* ermöglicht. Wie Ficino betont, ist die Kraft der Seele eine natürliche Kunst (*ars naturalis*) des Temperierens, eine

58) Der Begriff der *temperatio*, hier mit *Temperierung* übersetzt, beinhaltet zugleich das Moment der Regulation und dasjenige der ordnenden Organisation. Beides, Regulation und Organisation, muss bei *temperatio* bzw. *Temperierung* daher immer mitgedacht werden, um das Leistungsspektrum dieses Begriffs adäquat zu verstehen.

59) Marsilio Ficino, *Theologia Platonica – Platonic Theology*, Hg. James Hankins u. William Bowen, übersetzt v. Michael J. B. Allen u. John Warden, Bd. 2 (Cambridge/Mass./London, 2002), Liber IX, Kap. 6, Abs. 3, 176 – weiterhin wie folgt abgekürzt: IX, 6 (3), 176: "Illud vero vinculum libentius animam, immo virtutem animae appellabimus, siquidem animae virtus in corpore idem ferme est quod vel coaguli virtus in lacte quod inde coagulatur, vel temperatio artis in horologio quod inde ordine mirabili regitur, movetur et sonat. Ipsa vero anima non est virtus temperatiove humorum, sed forma quaedam substantialis, simplex, indivisibilis, quae virtute et arte naturali spiritus, humores membraque temperat." – Aufmerksam auf diese Stelle wurde ich durch den Aufsatz von Stéphane Toussaint, "Ficino," 316 f.

Kunst, die sich – durch menschliche Nachahmung oder Simulation der Natur[60] – im stundenschlagenden Horologium spiegelt ("temperatio artis in horologio"). Um aber die Analogie von Uhrwerk und Körper sowie von *temperatio artis* und *virtus animae* in all ihren Implikationen noch besser verstehen zu können, müssen zunächst die einzelnen Elemente dieser Analogie eingehender untersucht werden. Dazu gehören vor allem das *vinculum*, das Band, das nach Ficinos Definition die Kraft der Seele ist, die *temperatio*, die, weil für eben diese *virtus animae* stehend, als die sie in besonderer Weise auszeichnende Eigenschaft bezeichnet werden kann, und die *ars* selbst.

Funktion eines *vinculum* ist es, zu binden, zusammenzuhalten; das Binden und Zusammenhalten indes geschieht, so legt es der Text nahe, durch *temperatio*. Die Tätigkeit des Temperierens und bindenden Zusammenhaltens stellt Ficino als eine Kraft (*vis*, *virtus*) vor, die die Seele auf den Körper, genauer: auf die *spiritus*, die vier Körpersäfte (Blut, schwarze und gelbe Galle, Schleim) und die Gliedmaßen ausübt. Die natürliche, von der *anima* ausgeübte Kunst der temperierenden Regulation und Organisation sorgt folglich dafür, dass alle Teile des Körpers – *spiritus*, *humores*, *membra* – miteinander verknüpft, in einem bestimmten proportionalen Verhältnis geordnet sind und zusammengehalten werden. Die Annahme einer den Körper regulierenden seelischen *virtus* aber ist selbstverständlich nur dann nötig, wenn Körper und Materie nicht ihrerseits die autonome Fähigkeit der selbstregulativen Organisation besitzen, sondern, um überhaupt zu einer Einheit werden und als eine solche bestehen zu können, einer unkörperlichen einenden Substanz oder substanziellen Form bedürfen, die die Regulation und Organisation von Körper und Materie leistet. So war auch Ausgangspunkt des sechsten Buches der *Theologia Platonica* die Absicht Ficinos, diverse materialistische Seelenauffassungen zu widerlegen und das Verhältnis von Seele und Körper als eines zwischen tätiger Kraft einer immateriellen Substanz und passivem Rezeptionsvermögen eines materiellen Substrats auszuweisen.

Die Argumentation Ficinos bis zur zitierten Stelle, in der er die Uhr-Metapher einführt, ist denkbar einfach und deutet bereits auf das

---

60) Ebenda, IV, 1 (5), 252: "Proinde, si ars humana nihil aliud est aliud quam naturae imitatio quaedam [...]."

Prinzip des Lebens als entscheidendes, evidentes Kriterium zur Differenzierung von Körper und Seele hin. Körper zu sein, so Ficino, trifft sowohl auf Lebewesen als auch auf Nicht-Lebewesen zu. Wäre die vegetative und ernährende Kraft (*vis vegetativa et nutritiva*) nun selbst Materie oder Körper, dann würde nicht einzusehen sein, warum nicht auch Steine wachsen und sich ernähren sollten wie Pflanzen und Tiere.[61] Also muss das Prinzip des Lebens etwas sein, was das Körperliche übersteigt und zu ihm hinzukommt, um lebendige Körper von nicht-lebendigen Körpern unterscheidbar zu machen und vor diesen auszuzeichnen: eine nicht-körperliche, immaterielle verlebendigende Form (*forma vivificans*):

> Die verlebendigende Form selbst werden wir Seele nennen; jene Materie aber, der die Seele beigesellt ist, sollten wir, weil es nur durch die Seele geschieht, dass sie am Leben teilhat, weder als Seele noch als Teil der Seele, sondern als ein beseeltes Etwas bezeichnen.[62]

Wie bei Cusanus stehen nun auch bei Ficino die Begriffe des Lebens und der Bewegung in einem engen Zusammenhang. Im fünften Buch schreibt Ficino der Seele das Vermögen der inneren und äußeren Selbstbewegung zu. Die innere Bewegung definiert er als Leben, das sich die selbstbewegende Seele selbst zu geben vermag.[63] Da die perfekte Bewegung die Bewegung aus eigenem Antrieb (*motus spontaneus*) ist, sich diese wiederum in der Kreisbewegung in höchstem Maße realisiert findet, ist die spontane in sich geschlossene Bewegung das wesentliche Indikament und Zeichen inneren Lebens ("vitae interioris [...] signum")[64]. Wie bereits bei Cusanus steht daher auch bei Ficino die durch die Weltseele temperierte Zirkulation der Planeten und Himmelssphären für das Leben schlechthin,[65] das auf sympathetische

[61] Ebenda, VI, 4 (1), 156-158.

[62] Ebenda, VI, 4 (2), 158: "Formam ipsam vivificam vocabimus animam; materiam vero illam cui adiuncta est, quia inde vitae fit particeps, neque animam neque animae partem, sed animatum quiddam esse dicemus."

[63] Ebenda, V, 1 (2), 12: "Internus motus vita est; itaque sibimet vitam suppeditat."

[64] Ebenda, IV, 1 (21), 280.

[65] Ebenda, IV, 1 (21), 280: "Nulli ergo corpori vita magis intrinseca est quam mundi sphaeris."

gleichsam astralmagische Weise auf das irdische Leben einwirkt. Die Seele ist ihrer Gattung nach folglich „eigentümliches inneres Prinzip aller Tätigkeit und Bewegung“,[66] „sich selbst generierendes und hervorbringendes Leben“ (“vita se generans et producens”)[67], daher auch „Existenzprinzip des Lebewesens“ (“essendi principium est animali”)[68] sowie „Lenkerin und Bewegerin des Körpers“ (“rectrix motrixque corporis”)[69]. Sie ist nicht im Körper einbegriffen oder diesem in irgendeiner Form einverleibt, sondern umfasst und umgreift ihn, in allen seinen Teilen ständig als Ganze präsent, selbst und hält ihn als „Band der gesamten Natur“ (“vinculum naturae totius”)[70] in seinen Teilen zusammen,[71] indem sie ihn reguliert, organisiert, temperiert:

> Und würde nun jemand gesagt haben, dass die Himmel durch eine gewisse temperierend-regulierende Organisation bewegt würden, so wie einst die ehernen Sphären des Archimedes gedreht worden sind, würde er sich auch dazu veranlasst finden zu gestehen, dass diese temperierend-regulierende Organisation selbst nicht weniger substanziell, lebendig und vernünftig ist als diejenige, durch welche die Glieder der irdischen Lebewesen bewegt werden […].[72]

---

66) Ebenda, IV, 1 (21), 280: “[…] proprium internumque operationum motionumque principium […].”

67) Ebenda, VI, 12 (8), 192.

68) Ebenda, VI, 7 (1), 166: “[…] essendi principium est animali.”

69) Ebenda, VI, 7 (2), 168.

70) Ebenda, I, 1 (3), 16.

71) Ebenda, VI, 6 (3), 164: “Anima in corpore est non comprehensa, sed comprehendens.”; VI, 8 (2), 170: “Quamvis autem per haec appareat corpus natura sua nullo modo posse firmiter in unitate consistere, tamen corpus quodlibet aliquo modo in suis partibus [con]nectitur, cunctis in unam totius corporis copulam conspirantibus. Est igitur in corpore aliquid ultra corporis naturam quod ipsum unit, sistet et continet – unit, inquam, distantiam partium; sistit fluxum perennem; continet invicem per suam simplicem consonantiam dissonas corporis qualitates.” (“Auch wenn nun offensichtlich ist, dass ein Körper seiner Natur nach keineswegs von sich aus fest in seiner Einheit bestehen bleiben kann, so ist jeder Körper dennoch auf irgendeine Weise in allen seinen Teilen verknüpft, die gemeinsam zu einem Band des gesamten Körpers zusammenstimmen. Es gibt also irgendetwas im Körper jenseits seiner Natur, das ihn eint, bestehen lässt und zusammenhält – es überbrückt einend, sage ich, die Distanz der Teile; es macht den andauernden Fluss in sich beständig und quasi ruhend; es hält durch seine einfache Konsonanz die dissonierenden Qualitäten des Körpers zusammen.”).

72) Vgl. ebenda, IV, 1 (19), 278: “Ac si quis dixerit temperatione quadam moveri caelos, sicut aeneae quondam Archimedis volvebantur sphaerae, compelletur eam

Das temperierdend-organisierende Agieren geht – das wird bereits an dieser Stelle deutlich, aber schließlich im sechsten Buch der *Theologia Platonica* kurz vor der Einführung der Uhr-Metapher klar expliziert – von der „agentis substantia“ der Seele aus.[73] Die von Ficino in diesem Zusammenhang genannten ehernen Sphären des Archimedes, Teil einer von diesem entwickelten Maschine, die offenbar die *revolutiones* der Planeten zu simulieren vermochte, verweisen, ähnlich wie die bereits eingangs zitierte Uhr-Metapher, auf die die Natur nachahmende *ars* der *temperatio* – freilich nur implizit und ohne dass der Begriff der *ars* dabei verwendet wird.

Es stellt sich nun aber die Frage, was im Uhr-Modell Ficinos oder in der Sphärenmaschine des Archimedes eigentlich der regulativen Funktion der Seele entsprechen soll: Ist es tatsächlich die mechanische Uhrwerkhemmung selbst, so dass man von einem „pre-Cartesian spirit“ reden könnte, “where the power of the embodied soul (*virtus animae*) is compared to a mechanical balance ‘ruling, moving and making the clock chime’”?[74] Oder ist es doch eher jene die Teile der Uhr kunstvoll zusammensetzende Fertigkeit des Künstlers, eine technisch-künstlerische Organisation der Materie durch den Uhrmacher, dessen temperierende Tätigkeit schließlich im horologischen Kunstwerk allenfalls simulativ fortlebt und darin gespiegelt wird?

Selbst wenn von Ficino beabsichtigt gewesen wäre, die *virtus animae* mit der mechanischen Uhrwerkhemmung zu analogisieren, so muss doch betont werden, dass Descartes, anders als Ficino, die Uhrwerkhemmung nicht mit einem seelischen Vermögen, sondern mit dem selbstregulativen Mechanismus des Körpers in Bezug setzt. Während Ficino und Cusanus das Horologium als technische Simulation eines immateriellen, psychischen Bewegungsprinzips interpretieren, deutet Descartes die Uhr als Verkörperung eines rein physischen Mechanismus, der ohne die Wirkkraft einer immateriellen Substanz auskommt.

Es ist allerdings auch zweifelhaft, ob in Ficinos Uhr-Modell die Uhrwerkhemmung wirklich mit der *virtus animae*, den Körper in sich

---

ipsam temperationem non minus substantialem, vitalem, rationalem esse fateri quam sit temperatio qua terrenorum animalium membra moventur [...].”

73) Ebenda, VI, 9 (3), 176: “Vis actioque agenti alicui intima sine agentis substantia non existit. In illis ergo est anima, si in illis viget et agit.”

74) Toussaint, “Ficino,” 316-317.

'binden', vereinen, zusammenhalten und ihm Beständigkeit geben zu können, zusammengedacht und gleichgesetzt wird. Darüber, wie die Uhr-Metapher bei Ficino zu verstehen sein könnte, geben hingegen folgende Definitionen von *ars* und *natura* aus dem vierten Buch der *Theologia Platonica* Aufschluss. Hier heißt es: „Was ist menschliche Kunst? Eine gewisse Natur, die die Materie von außen bearbeitet. Was ist Natur? Eine Kunst, die die Materie von innen temperierend organisiert, etwa so, als wäre ein Zimmermann im Holz anwesend."[75]

Wenn Ficino also in der Uhr-Metapher von der „temperatio artis in horologio" spricht, dann scheint er damit zu verstehen geben zu wollen, dass die Seele, die als organisierende Wirkkraft der Natur in den *res naturales* selbst tätig ist, im horologischen Kunstwerk nur simulativ, dem Anschein nach und folglich nur repräsentativ als Resultat der die Materie von außen imprägnierenden, bearbeitenden und organisierenden *virtus* des menschlichen *artifex* anwesend ist. Zwischen Natur und Kunst, zwischen natürlicher und künstlicher regulierender Organisation besteht demnach eine erhebliche Differenz: Der Schöpfung ist die einende, verlebendigende Kraft der *anima* überall *von innen her* gegenwärtig; in der Kunst hingegen wird diese Kraft vom *artifex* auf das Kunstwerk *von außen* transferiert, so dass das Kunstwerk nur solange die auf es übertragene Seelenbewegung fortsetzt, wie es selbst Bestand hat oder bis es von außen nachreguliert wird. Gerade der horologische Automat aber repräsentiert die partizipative Annäherungsbewegung der menschlichen Kunst und Technik an die animierte Natur auf besonders vollkommene Weise, weil es die perfekte, spontane Seelenbewegung, den Zirkel, und damit die nach Maß, Zahl und Gewicht organisierte Lebendigkeit der Schöpfung nachzustellen vermag. Das Horologium wird dadurch bei Ficino sowohl zum physikotheologischen Signum der sich über die Natur bis hin zur Kunstfertigkeit des Menschen erstreckenden schöpferischen Kraft Gottes, als auch zum explikativen, veranschaulichenden Transfermodell, das auf die temperierende, organisierende, regulierende, Leben überhaupt erst realisierende Tätigkeit des substanziellen Seelenagenten übertragen und appliziert werden kann.

---

75) Ficino, *Theologia Platonica*, IV, 1 (5), 252: "Quid est ars humana? Natura quaedam materiam tractans extrinsecus. Quid natura? Ars intrinsecus materiam temperans, ac si faber lignarius esset in ligno."

## 5. Descartes' Uhr-Metapher als Modell für die Mechanik des Unwillkürlichen

Im fünften naturphilosophischen Teil seines 1637 erschienenen *Discours de la méthode* unternimmt Descartes ein auf experimentelle Erfahrung hinzielendes und zugleich auf dieser basierendes Gedankenexperiment, um das Wesen der materiellen Dinge ("la nature des choses matérielles")[76] zu ergründen und die Gesetze zu beschreiben, nach denen diese geordnet und organisiert sind, ohne aber substanzielle Formen oder eine vernünftige, vegetative oder sensitive Seele vorauszusetzen.[77] Descartes versucht nachzuweisen, dass der organische Körper keiner unmittelbaren Beteiligung der Seele oder des Bewusstseins bedarf, um seine Tätigkeiten zu verrichten. So stellt er am Beispiel der Herz- und Arterienbewegung ("mouvement du cœur et des artères")[78] dar, wie der lebende Körper allein nach den ihm inhärenten mechanischen Gesetzen (*règles des mécaniques*) autonom zu funktionieren vermag.[79] Nachdem er die Mechanik des Herzens und der Arterien beschrieben hat, führt er in einem eingeschobenen Resümée seiner Methode und Argumentation noch vor der Schilderung des Blutkreislaufs die Uhr-Metapher ein:

> Damit übrigens diejenigen, die die Überzeugungskraft mathematischer Beweise nicht kennen und es nicht gewohnt sind, wahre Beweisgründe von bloß wahrscheinlichen zu unterscheiden, dieser meiner Erklärung nicht auf gut Glück und ohne Prüfung widersprechen, so mögen sie sich gesagt sein lassen, daß der soeben erklärte Mechanismus [*ce mouvement, que je viens d'expliquer*] sich allein aus der Einrichtung [*disposition*] der Organe ergibt, die man im Herzen mit seinen Augen sehen, aus der Wärme, die man dort mit seinen Fingern spüren, und aus der Natur des Blutes, die man durch Erfahrung kennenlernen kann, und dies mit der gleichen Notwendigkeit, wie der Mechanismus einer Uhr aus der Kraft, Lage und Gestalt ihrer Gewichte und Räder folgt.[80]

---

76) René Descartes, *Discours de la méthode pour bien conduire sa raison, et chercher la verité dans les sciences*, in ders., *Philosophische Schriften* (Hamburg, 1996), 68.

77) Ebenda, 74. Zum Zusammenhang vgl. Dennis Des Chene, *Spirits and Clocks: Machine and Organism in Descartes* (Ithaca, 2001), dort auch weiterführende Literatur.

78) Descartes, *Discours*, 76.

79) Ebenda, 88.

80) Ebenda, 80 f.: "Au reste, afin que ceux qui ne connaissent pas la force des démonstrations mathématiques, et non sont pas accoutumés à distinguer les vraies raisons des

Der von Descartes bemühte Vergleich zwischen Horologium und Herz-Blut-Bewegung impliziert zwei grundsätzliche Aspekte: Zum einen entspricht der Disposition der körpereigenen Organe (hier Herz und Blutgefäße) die Anordnung und Beschaffenheit der das Uhrwerk konstituierenden Teile, zum anderen aber ist beiden gemeinsam, dass erst aus der spezifischen Disposition und Konstitution der Teile ihre bestimmte, gleichsam auf einen einzigen singulären Zweck fixierte und beschränkte Tätigkeit folgt. Dabei ist der Zweck nicht etwas, von dem Descartes annähme, dass er den kooperierenden Bewegungsfolgen der Teile *per se* schon zukäme; sondern die organische Disposition zu bestimmten Bewegungsfolgen reicht für sich genommen vollständig aus, die Tätigkeit des Körpers – quasi die Resultante aus der Koordination der Teile – zu erklären, ohne von einer dahinter stehenden willkürlichen Intention auszugehen, die man ihr allenfalls im Nachhinein als Zweck oder Aufgabe zusprechen könnte.

Die Wahl der Uhr als Vergleichsmodell für die Herz-Blut-Bewegung findet ihre zusätzliche Begründung in der gleich im folgenden Abschnitt von Descartes dargelegten Mechanik des Blutkreislaufs. Denn da Descartes die beständige Zirkulation ("circulation perpétuelle")[81] des Blutes unmittelbar nach Einführung der Uhr-Metapher expliziert, lässt das horologische Modell sich nun auch auf die Kreisbewegung des Blutes applizieren. Demzufolge funktioniert der Blutkreislauf genauso autonom-spontan (*sua sponte*) wie der horologische Automat, sobald dieser einmal eingerichtet und in Bewegung gesetzt ist. Descartes' Experiment, die Schöpfung des Menschen durch Gott gedanklich zu simulieren, indem er allein das sinnlich Konstatierbare, die Materie, als Konstituens für Bewegung und Leben annimmt, bedingt somit auch die Analogie von horologischem und göttlichem *mechanicus*. Ähnlich wie der Uhrmacher die materiellen Bestandteile der Uhr so disponiert hat, dass

---

vraisemblables, ne se hasardent pas de nier ceci sans l'examiner, je les veux avertir que ce mouvement, que je viens d'expliquer, suit aussi nécessairement de la seule disposition des organes qu'on peut voir à l'œil dans le cœur, et de la chaleur qu'on y peut sentir avec des doigts, et de la nature du sang qu'on peut connaître par expérience, que fait celui d'un horologe, de la force, de la situation et de la figure de ses contrepoids et de ses roues." – Die zitierte deutsche Übersetzung des *Discours* stammt hier und im Folgenden von L. Gäbe.

81) Ebenda, 82.

die Uhr ohne sein eigenes Zutun von selbst und aus eigenem Antrieb zu funktionieren vermag und ihren Zweck der Zeitindikation erfüllt, hat auch Gott, freilich ungleich perfekter, die materiellen Elemente des lebendigen Körpers so eingerichtet, dass dies alleine hinreicht, um Bewegung und Leben zu ermöglichen.[82] Der Disposition der einzelnen Organe entspricht dabei jedoch nur eine bestimmte verrichtende Tätigkeit resp. Funktion, wie auch die Uhr bis auf die Funktion der Zeitbestimmung keine zusätzliche Aufgabe hat. Genausowenig aber wie das Horologium *willkürlich* die Zeit anzeigt, bewegt sich der lebendige Körper willentlich, *per voluntatem*. So ist es das unwillkürliche Moment der notwendig aus der Disposition der Teile folgenden animalischen Bewegung, das bei Descartes sowohl den tierischen als auch den menschlichen Körper bestimmt. Um nun den Unterschied zwischen Tier und Mensch zu illustrieren, vergleicht Descartes artifizielle Automaten-Menschen, die menschliche Handlungen so weit wie möglich simulieren, und sogenannte „wahre Menschen" ("vrais hommes")[83]. Während Maschinen-Menschen Bewegungen nur infolge der speziellen Einrichtung ihrer Teile auszuführen vermögen und somit die Anzahl ihrer Handlungsoptionen begrenzt ist, sie also meist nur eine einzige (mechanisch programmierte) Bewegungsabfolge nachahmen, zeichnet sich der Körper „wahrer Menschen" dadurch aus, dass mit ihm ein deutlich höheres Handlungspotenzial realisierbar ist. Dazu zählt

---

82) Vgl. ebenda, 90: "Ce qui ne semblera nullement étrange à ceux qui, sachant combien de divers *automates*, ou machines mouvantes, l'industrie des hommes peut faire, sans y employer que fort peu de pièces, à comparaison de la grande multitude des os, des muscles, des nerfs, des artères, des veines, et de toutes les autres parties qui sont dans le corps de chaque animal, considéront ce corps comme une machine, qui, ayant été faites des mains de Dieu, est incomparablement mieux ordonnée, et a en soi des mouvements plus admirables, qu'aucune de celles qui peuvent être inventées par les hommes." ("Dies wird dem keineswegs sonderbar vorkommen, der weiß, wie viele verschiedene *Automaten* oder bewegungsfähige Maschinen menschliche Geschicklichkeit zustandebringen kann, und dies unter Verwendung nur sehr weniger Einzelteile verglichen mit der großen Anzahl von Knochen, Muskeln, Nerven, Arterien, Venen und all den anderen Bestandteilen, die sich im Leibe jedes Tieres finden. Er wird diesen Leib für eine Maschine ansehen, die aus den Händen Gottes kommt und daher unvergleichlich besser konstruiert ist und weit wunderbarere Getriebe in sich birgt als jede Maschine, die der Mensch erfinden kann.").

83) Ebenda, 92.

Descartes vor allem die Sprache, die bei Tieren, selbst wenn deren Stimmapparat dem menschlichen mitunter sehr nahe kommt, zumindest in dieser Ausdrucksfülle und Komplexität nicht anzutreffen ist, sowie, als zweites Unterscheidungsmerkmal, Handlungen, die aus vernünftiger Einsicht (*connaissance*) erfolgen:

> Denn die Vernunft ist ein Universalinstrument, das bei allen Gelegenheiten zu Diensten steht, während diese Organe für jede besondere Handlung einer besonderen Einrichtung bedürfen; was es unwahrscheinlich macht, daß es in einer einzigen Maschine genügend verschiedene Organe gibt, die sie in allen Lebensfällen so handeln ließen, wie uns unsere Vernunft handeln läßt.[84]

Die im Einzelfall zu einer gewissen Überlegenheit über den Menschen führende organische und anatomische Spezialisierung der Tiere wird

---

[84] Ebenda, 92 f.: "Car, au lieu que la raison est un instrument universel, qui peut servir en toutes sortes de rencontres, ces organes ont besoin de quelque particulière disposition pour chaque action particulière; d'où vient qu'il est moralement impossible qu'il y en ait assez de divers en une machine pour la faire agir en toutes les occurrences de la vie, de même façon que notre raison nous fait agir." – Unter vernünftigen Handlungen dürfte Descartes moralische und wissenschaftliche Tätigkeiten verstehen, sicher aber auch solche, die einem spirituellen Habitus, wie etwa der Habitus des Glaubens, der Gnade, der Frömmigkeit etc., entspringen. Vgl. dazu René Descartes, *Œuvres philosophiques de Descartes,* Tôme IV, *Les lettres philosophiques*, Hg. Adolphe Garnier (Paris, 1835), 27 (Lettre XXXV, Brief an Regius): "Nous ne nions pas aussi les habitudes, mais nous les comprenons sous un double genre: les unes purement matérielles, qui dépendent de la seule configuration, ou autre disposition des parties; et les autres immateriélles ou spirituelles, comme les habitudes de la foi, de la grace [...]." ("Wir verneinen auch nicht die Gewohnheitshandlungen, fassen sie aber in doppelter Art: die einen als rein materielle, die von der alleinigen Konfiguration oder sonstigen Disposition der Teile abhängen; und die anderen als immaterielle oder geistige, wie die Gewohnheitshandlungen des Glaubens, der Gnade [...]."). – Zur Veranschaulichung des Unterschieds zwischen künstlichen und natürlichen Automaten benutzt Descartes auch hier die Uhr-Metapher, ebd., 28: "Cette différence pourtant du plus ou du moins est grande, comme vous dites, parce que le peu de roues qui composent une horloge ne peuvent entrer en aucune comparaison avec le nombre infini d'os et de nerfs, de veines, d'artères, etc. qui se trouvent dans le plus vil de tous les plus petits animaux." ("Dieser Unterschied ist jedoch mehr oder weniger groß, wie Ihr sagt, da die geringe Anzahl an Rädern, aus denen sich eine Uhr zusammensetzt, in keinster Weise den Vergleich mit der unendlichen Anzahl an Knochen und Nerven, an Venen, Arterien etc. antreten kann, die sich selbst in den geringsten der kleinsten Lebewesen findet.").

von Descartes folglich von der sich auf die für willkürliche Handlungen nötige Bewegung des Körpers auswirkenden Universalität des menschlichen Bewusstseins deutlich unterschieden. Zur Veranschaulichung dieser Differenz rekurriert Descartes erneut auf das Beispiel der Uhr:

> Aber sie [*die Tiere*] haben im Gegenteil gar keinen [*Geist*; *esprit*], und es ist die Natur, die in ihnen je nach der Einrichtung ihrer Organe wirkt, ebenso wie offensichtlich eine Uhr, die nur aus Rädern und Federn gebaut ist, genauer die Stunden zählen und die Zeit messen kann als wir mit all unserer Klugheit [*prudence*].[85]

Das horologische Modell steht bei Descartes also – im Unterschied zu Cusanus' und Ficinos Gebrauch der Uhr-Metapher – ausschließlich für ein in sich geschlossenes, *unbeseeltes* körperliches System,[86] dessen Bewegungen aus der Anordnung und Konstitution seiner organischen Teile und den Gesetzen der Mechanik resultieren. In der sechsten *Meditatio* seiner 1641 in Paris erschienenen *Meditationes de prima philosophia* stellt Descartes, wiederum am Beispiel der Uhr, das besondere Kennzeichen dieser Bewegungen heraus: Sie finden, so Descartes, unabhängig von der „Herrschaft des Willens" (*non ab imperio voluntatis procedunt*) und der Existenz des Geistes allein aus naturgesetzlicher Notwendigkeit statt. Das aber heißt, dass die von Descartes als methodisches (Kontrast)Modell konzipierte natürliche Körper-Maschine eine reine mechanische Organik des Unwillkürlichen vorstellt, die sich durch den seelenlosen horologischen Automaten angemessen versinnbildlichen lässt, auch wenn artifiziell konstruierte Apparate in Komplexität und Perfektion weit hinter natürlichen Automaten zurückstehen. Descartes fasst diesen Zusammenhang in seiner 1664 posthum veröffentlichten Schrift über die Formation des Fötus prägnant zusammen:

---

85) Descartes, *Discours*, 96 f.: "[...] mais plutôt qu'ils n'en ont point, et que c'est la nature qui agit en eux, selon la disposition de leurs organes: ainsi qu'on voit qu'un horologe, qui n'est composé que de roues et de ressorts, peut compter les heures, et mesurer le temps, plus justement que nous avec toute notre prudence."

86) *Unbeseelt* bezieht sich auf einen Seelen-Begriff, der von der Vorstellung eines unmittelbaren, kausalen Einflusses einer vegetativen und sensitiven Seele auf körperliche Tätigkeiten, Verrichtungen und Bewegungen ausgeht.

> Aus diesem Grund habe ich mich bemüht, hier die ganze Maschine unseres Körpers auf eine solche Weise zu beweisen und zu erklären, dass wir uns genausowenig veranlasst sähen zu denken, unsere Seele rufe in uns die Bewegungen hervor, von denen wir keineswegs erfahren, dass sie durch unseren Willen gelenkt werden, wie wir Anlass haben zu urteilen, dass es eine Seele in der Uhr gibt, die sie die Stunden anzeigen lässt.[87]

Über den sich in unwillkürlichen Bewegungen äußernden mechanisch regulierten Körper-Uhr-Automaten ohne *esprit* oder *mens* hinaus ist aber der menschliche Körper auch exekutives Instrument willkürlicher, vom Bewusstsein veranlasster bzw. indirekt gesteuerter Bewegungen. Das ändert jedoch nichts daran, dass auch intentionale, willkürliche Bewegungen nur dann realisierbar sind, wenn die mechanische Disposition der Organe und Gliedmaßen dies zulässt.[88]

---

87) René Descartes, *L'Homme de René Descartes et un traitté de la formation du foetus du mesme autheur* (Paris, 1664), 112: "[...] c'est pourquoy ie tascheray icy de le prouuer, & d'expliquer tellement toute la Machine de nostre Corps, que nous n'aurons pas plus de sujet de penser que c'est nostre Ame qui excite en luy les mouuemens que nous n'experimentons point estre conduits par nostre volonté, que nous en auons de iuger qu'il y a vne Ame dans vne horloge, qui fait qu'elle monstre les heures."

88) Vgl. dazu Descartes, *L'Homme*, 111: "Ce qui montre que l'Ame ne peut exciter aucun mouuement dans le Corps, si ce n'est que tous les organes corporels, qui sont requis à ce mouuement, soient bien disposez: mais que tout au contraire, lors que le Corps a tous ses organes disposez à quelque mouuement, il n'a pas besoin de l'Ame pour le produire; Et que par consequent tous les mouuemens que nous n'experimentons point dépendre de nostre pensée, ne doiuent pas estre attribuez à l'Ame, mais à la seule disposition des organes; Et que mesme les mouuemens, qu'on nomme *voluntaires*, procedent principalement de cette disposition des organes, puis qu'ils ne peuuent estre excitez sans elle, quelque volonté que nous en ayons, bien que ce soit l'Ame qui les determine." ("Was zeigt, dass die Seele keine einzige Bewegung im Körper hervorrufen kann, wenn nicht alle körperlichen Organe, die für diese Bewegung erforderlich sind, entsprechend eingerichtet sind. Vielmehr ist ganz das Gegenteil der Fall: Sobald der Körper alle Organe hat, die zu einer bestimmten Bewegung disponiert sind, bedarf er keineswegs der Seele, um sie [die Bewegung] zu erzeugen. Was folglich auch zeigt, dass alle Bewegungen, von denen wir keineswegs erfahren, dass sie von unserem Denken abhängen, nicht der Seele, sondern allein der Disposition der Organe zugesprochen werden dürfen; ja, dass sogar die Bewegungen, die man willkürliche nennt, im Wesentlichen aus eben dieser Disposition der Organe hervorgehen, da sie ohne diese nicht hervorgerufen werden können, egal, welchen Willen wir hierin haben mögen, auch wenn es die Seele ist, die sie [die willkürlichen Bewegungen] bestimmt.").

Die Konstanz, mit der Descartes über Jahrzehnte hinweg am horologischen Vergleichs- und Transfermodell als eines seelen- und willenlosen Automaten zur Veranschaulichung der körperlichen Mechanik des Unwillkürlichen festgehalten hat, zeigt, wie wichtig ihm die Uhr als prototypische *machina machinarum* gewesen ist. Historisch markiert seine Verwendung der Uhr-Metapher zugleich eine deutliche Verschiebung ihrer Funktion: War sie bis dahin ein repräsentatives, technomorphes Signum für die Seele als Prinzip der (Selbst)Bewegung und des Lebens, so machte sie Descartes zum Sinnbild einer ausschließlich nach natürlichen Gesetzmäßigkeiten der Materie funktionierenden Mechanik des lebendigen Körpers und seiner unwillkürlichen Bewegungen.

## 6. Seelenmaschine, Körpermaschine: Die Uhr-Metapher bei Leibniz

Leibniz nutzt, anders als Descartes, die Uhr-Metapher nicht in einem einheitlichen und konsistenten Sinn. In seinen Schriften findet sich neben der naturtheologischen Konnotation des „horologium Dei" oder des „horologium mundi" als Sinnbild für die von Gott eingerichtete „schöne Ordnung"[89] der Schöpfung, „in der alles durch eine bestimmte Notwendigkeit auf die höchste Zusammenstimmung aller Dinge führt",[90] der Uhr-Begriff auch im Kontext von Überlegungen, die an diejenigen von Cusanus erinnern. Daran schließen sich zentrale Reflexionen über die an Zahlenreihen gemahnende Logik der „Serie der Dinge" an, in der es stets „eine bestimmte Verknüpfung" gibt, „auch wenn sie nicht immer notwendig ist".[91] Hierbei steht das horologische

---

[89] Brief an Jakob Thomasius vom 30. April 1669, in Gottfried Wilhelm Leibniz, *Die Philosophischen Schriften von Gottfried Wilhelm Leibniz*, Hg. Carl Immanuel Gerhardt, 7 Bde (Hildesheim/New York, 1978; Nachdruck der 1. Aufl. 1875-1890), Bd. I, 25: "Ordo vero pulcher ex eo oriatur, quia est horologium Dei." – Weiterhin zitiert als GP (hier z. B. GP I, 25).

[90] Brief an Jakob Thomasius vom 29. Dezember 1670, GP I, 33: "[...] ita instituisse horologium mundi, ut cuncta deinde, velut necessitate quadam ad summam omnium harmoniam consequerentur."

[91] Gottfried Wilhelm Leibniz, *Divisio terminorum ac enumeratio attributorum* (zw. 1687/1690), in ders., *Sämtliche Schriften und Briefe*, Hg. Berlin-Brandenburgische Akademie der Wissenschaften, Reihe VI: Philosophische Schriften, Bd. 4, Hg. Leibniz-

Vergleichsmodell für den Zusammenhang von Zeit resp. Sukzession und der kausalen, wenn auch nicht immer notwendigen Abfolge von *dinglichen* Zuständen (*status*, *état*), die in letzter Konsequenz auf die *prima causa* verweisen:

> [...] denn die Uhren bewirken nicht, sondern indizieren nur das Vorher und Nachher, das Früher und Später. Das gleiche gilt für die Bewegung des Himmels: Sie unterscheidet sich nämlich nicht von der Bewegung der Uhr [...]. Die Wurzel [*radix*] aber der Zeit ist die erste Ursache [*causa prima*], die die Abfolgen der Dinge virtuell in sich enthält und macht, dass alles zugleich oder früher oder später ist.[92]

Die virtuelle oder dynamische Einfaltung der Abfolgen der Dinge in der ersten Ursache besteht in der – freilich zeitlosen – mathematischen Konzeption der Schöpfung im Geist Gottes. Die Uhr versinnbildlicht bei Leibniz aber nicht mehr wie bei Cusanus das Verhältnis von göttlicher Ewigkeit und geschöpflicher Zeit, in dem Gott trotz seiner Transzendenz in allem Geschaffenem gegenwärtig ist. Vielmehr veranschaulicht sie nur noch die Zeit als Prinzip der Sukzession selbst, deren Messinstrument und Indikator sie ist. Da Leibniz die Zeit als „imaginäres Seiendes" definiert,[93] wird folglich auch die Uhr zu einem repräsentativen Instrument des Imaginären, d. h. des messenden menschlichen Geistes, der sich, so Leibniz in einem frühen Brief an Jakob Thomasius, „nichts anderes vorstellen kann als Geist (sobald dieser sich nämlich selbst denkt), Raum, Materie, Bewegung und dasjenige, was aus deren Kombination resultiert."[94] Damit erhält die Uhr

Forschungsstelle der Universität Münster, Teil A, bearbeitet von Heinrich Schepers (Berlin, 1999), 563; weiterhin abgekürzt wie folgt: AA VI 4A, 563: "Idemque est in alia quavis serie rerum, semper enim certa connexio est, etsi necessaria semper non sit."

92) Leibniz, *Definitiones notionum metaphysicarum atque logicarum* (zw. 1687/1690), AA VI 4A, 629: "[...] nam horologia non efficiunt sed indicant tantum prioritatem et posterioritatem. Idem est de motu coeli neque enim a motu horologii differt [...] Radix autem temporis est in causa prima, successiones rerum virtute in se continente, quae facit, ut omnia sint simul, aut priora vel posteria."

93) Leibniz, *Definitiones*, AA VI 4A, 629: "Tempus est Ens imaginarium [...]."

94) Brief an Jakob Thomasius vom 30. April 1669, GP I, 24-25: "Iam vero mens humana nihil aliud imaginari potest, quam mentem (quando scilicet cogitat se ipsam), spatium, materiam, motum et quae ex his inter se comparatis resultant; quidquid

einen indirekten Bezug zur göttlichen „Wurzel" ("radix") der Zeit, die, weil sie reale Bedingung alles imaginär Zeitlichen ist, zugleich die Bedingung jedes zeitlichen Indikators wie der Planetenbewegung und des diese nach menschlicher Kunstfertigkeit und Wissenschaft eingerichteten messenden und simulierenden Horologium vorstellt. Darüber hinaus führt das horologische Getriebe vor Augen, dass die Sukzessivität von Bewegungen, wie derjenigen des Stundenzeigers, in der ineinandergreifenden Mechanik der Dinge, hier der „Räder", d. h. in den sich auseinander ergebenden momentanen Zuständen der Dinge, in diesem Fall der Räder des Uhrwerks, begründet sind:

> Freilich, von zwei kontradiktorischen Zuständen der gleichen Sache ist derjenige Zustand zeitlich früher, der der Natur nach früher ist, bzw. derjenige, der den Grund des anderen enthält. [...] Wie es bei einer Uhr erforderlich ist, dass wir, um den gegenwärtigen Zustand ihrer Räder auf vollkommene Weise zu verstehen, den Grund dieses gegenwärtigen Zustands begreifen, der im vorherigen Zustand enthalten ist. Dasselbe gilt für jede andere beliebige Serie von Dingen [...].[95]

Die gleichzeitige Auseinanderlegung resp. Entfaltung eines Zustands aus einem vorherigen Zustand und deren erneute logische Verknüpfung ist, folgt man der Argumentation von Leibniz, allein ein rationaler und damit ein weitestgehend imaginärer Akt des menschlichen Geistes.

Für die weitere Verwendung der Uhr-Metapher bei Leibniz bildet nun die Aufeinanderfolge von Zuständen die passende Folie. Anders wäre nicht zu verstehen, warum Leibniz die Uhr einerseits für ein Sammelding (*assemblage*) ähnlich einer Materiemasse (*masse de la matière*) ansieht,[96] sie andererseits aber dennoch als Metapher für die

---

superaddideris, verba sunt tantum, quae nominari et inter se varie combinari possunt, explicari et intelligi non possunt."

95) Leibniz, *Divisio*, AA VI 4A, 563: "Porro ex duobus statibus contradictoriis ejusdem rei, is prior tempore est, qui natura prior est, seu qui alterius rationem involvit, [...]. Ut in horologio, ut perfecte intelligamus statum praesentem rotarum, requiritur [ut] intelligamus rationem ejus, quae continetur in statu praecedente; atque ita porro. Idemque est in alia quavis serie rerum [...]."

96) Vgl. Gottfried Wilhelm Leibniz, *Systeme nouveau de la nature et de la communication des substances, aussi bien que de l'union qu'il y a entre l'ame et le corps*, in ders., *Philosophische Schriften*, Hg. Hans Heinz Holz, Bd. 1 (Frankfurt/Main, 1996), 214; ders., *Eclaircissement du nouveau systeme de la communication des substances, pour servir*

Seelen-Substanz gebraucht, die er im *Système nouveau de la nature* von 1695 auch als „geistigen oder formellen Automaten" ("Automate spirituel ou formel") bezeichnet.[97]

So führt er das Zwei-Uhren-Gleichnis im *Second Éclaircissement du Système de la Communication des Substances* ein, um seine Hypothese der präetablierten Harmonie von Seele und Körper gegen die Theorien des wechselseitigen psychophysischen Einflusses und des Okkasionalismus abzusetzen. Beide Uhren repräsentieren voneinander unabhängige Abfolgen von Zuständen, die unterschiedlichen Gesetzmäßigkeiten gehorchen und durch einen „zuvorkommenden Kunstgriff" ("artifice prevenant") Gottes von Beginn an aufeinander abgestimmt wurden.[98]

Das Prinzip der Abfolge von Zuständen als die gemeinsame Folie von Körper und Seele wird dann von Leibniz in den 1705 publizierten *Considérations sur les principes de vie et sur les natures plastiques* explizit in den Vordergrund geholt.

Ausgangspunkt der *Considérations* ist zunächst Leibniz' Kritik an Descartes. Dieser habe zwar richtig erkannt, dass die Gesamtkraft der Körper in der Welt erhalten bleibe, die Kraft eines Körpers somit nicht von der Seele beeinflusst werden könne; er habe hingegen noch nicht realisiert, dass auch die Bewegungsrichtung des Körpers letztlich der Quantität nach gleich bleibe, und sei deswegen davon überzeugt gewesen, die Richtung des Körpers könne durch die Seele indirekt über die Animalgeister gesteuert werden. Im Gegensatz zu Descartes sieht sich

---

*de reponse à ce qui est dit dans le Journal du 12 Septembre 1695*, in ders., *Philosophische Schriften*, 228.

97) Leibniz, *Systeme nouveau*, 220.

98) Leibniz, *Second Éclaircissement du Système de la Communication des Substances*, in ders., *Philosophische Schriften*, 240. Das Zwei-Uhren-Gleichnis findet sich ebd., 238-243. – Zur Diskussion dieser Stelle sowie anderer Stellen, wo dieses Gleichnis verwendet wird, vgl. David Scott, "Leibniz and the Two Clocks," *Journal of the History of Ideas*, 58 (1997), 445-463. David Scott verkennt meiner Ansicht nach aber die Parallele zwischen der an Zahlenreihen erinnernden Abfolge von körperlichen und seelischen resp. monadischen Zuständen und damit die grundsätzliche Möglichkeit, die Uhr-Metapher sowohl für die Seele als auch für den Körper einsetzen zu können. Er verweist aber zurecht auf die Differenz zwischen der Auffassung des Okkasionalismus, wie sie bei Leibniz anzutreffen ist, und derjenigen, auf die sich Leibniz damit bezieht, nämlich Malebranches Okkasionalismus.

Leibniz aufgrund der beiden Naturgesetze von der quantitativen Erhaltung der Kraft/Bewegung und der Bewegungsrichtung dazu veranlasst, den Leib-Seele-Dualismus Descartes' noch zu radikalisieren, indem er Seele und Leib zwei voneinander separierten unterschiedlichen Gesetzmäßigkeiten folgen lässt, die jede für sich in präzise mathematische Sukzessionen transferierbar sind, aber in keinerlei wirk- oder gelegenheitsursächlichem Verhältnis zueinander stehen:

> Die Seelen folgen ihren eigenen Gesetzen, die in einer bestimmten Entwicklung ihrer Vorstellungen gemäß dem Guten und Bösen bestehen, während die Körper ihrerseits ebenfalls den ihrigen, nämlich den Regeln der Bewegung, folgen. Trotzdem treffen diese beiden Wesenheiten von gänzlich verschiedener Art zusammen und entsprechen einander wie zwei Uhren, die vollkommen in derselben Weise reguliert worden sind [*deux pendules parfaitement bien reglées sur le même pied*], wenngleich sie vielleicht von gänzlich verschiedenem Bau sind.[99]

Sprach Leibniz im *Système nouveau* noch von zwei *horloges*, so fällt auf, dass Leibniz hier nun den Ausdruck *pendule* benutzt. Was aber könnte Leibniz dazu motiviert haben, von *pendule* statt von *horloge* zu sprechen? – Ich nehme an, dass Leibniz an dieser Stelle ausdrücklich auf die von Huygens entwickelte Pendeluhr rekurriert, die seinerzeit als eines der höchstentwickelten Präzisionsinstrumente und verlässlichsten Zeitmessgeräte galt. Leibniz nutzt also die *pendule* als Gleichnis für die mathematisch präzise Abfolge in den Zuständen von Seelen und Körpern sowie für die unüberbietbare Präzision Gottes in der einmaligen schöpferischen Justierung der *deux pendules* aufeinander, die schließlich, auf das Beste reguliert, „sur le même pied", auf dem gleichen Stand, funktionieren. Dies betrifft jedoch nicht nur die Übereinstimmung von Seele und Körper, zwischen finalursächlich regulierten monadischen Perzeptionen und nach mechanischen Gesetzen resp. kausalursächlich

---

[99] GP VI, 541: "Les Ames suivent leur loix, qui consistent dans un certain developpement des perceptions selon les biens et les maux; et les corps suivent aussi les leurs, qui consistent dans les regles du mouvement: et cependant ces deux Estres d'un genre tout à fait different, se rencontrent ensemble et se repondent comme deux pendules parfaitement bien reglées sur le même pied, quoyque peutestre d'une construction toute differente." – Die Übersetzung stammt aus Gottfried Wilhelm Leibniz, *Philosophische Werke in vier Bänden*, Bd. 1: *Hauptschriften zur Grundlegung der Philosophie*, Hg. Ernst Cassirer, übersetzt von Arthur Buchenau (Hamburg, 1996), Teil 1, 319.

bestimmten physischen Bewegungen, sondern auch die Abstimmung in der jeweiligen Abfolge der Bewegungen innerhalb des Körpers sowie der Perzeptionen innerhalb der Seele selbst. In Bezug auf die präetablierte Abstimmung zwischen Seele und Körper schreibt Leibniz: „Denn dieses allgemeine Prinzip [...] führt indes [...] um so sicherer und klarer zu dem ersten allgemeinen Beweger, von dem in gleicher Weise der Zusammenhang [besser: die Abfolge, da frz. *la suite*] und die Übereinstimmung aller Vorstellungen kommt.“[100]

Was aber meint nun „suite“ (Abfolge, Zusammenhang) genau? – Bereits im *Eclaircissement des difficultés* von 1696 hatte Leibniz auf die „Genauigkeit“ (“exactitude”) als Kriterium für das Potenzial der *pendule*-Metapher hingewiesen, die Seele in der Kontinuität und Sukzession ihrer Perzeptionen versinnbildlichen zu können:

> Ich habe die Seele nur hinsichtlich der geregelten Genauigkeit der Veränderungen [*l'exactitude reglée des changemens*] mit einer Pendeluhr verglichen, und diese Genauigkeit ist selbst bei den besten Uhren nur unvollkommen, bei den Werken Gottes hingegen vollkommen; und man kann sagen, daß die Seele ein sehr genauer immaterieller Automat ist.[101]

Da der Wechsel der Zustände jeder Substanz, also auch jeder Seele, einem individuellen Ordnungsgesetz folgt, so dass jeder perzeptive Zustand aus den vergangenen perzeptiven Zuständen hervorgeht und in jedem momentanen Zustand die zukünftigen Zustände schon eingefaltet sind,[102] besteht, so Leibniz im gleich auf das *pendule*-Gleichnis

---

[100] Leibniz, *Hauptschriften zur Grundlegung der Philosophie*, 320. – GP VI, 542: “Ce principe general [*die präetablierte Harmonie*] [...] nous mene d'autant plus seurement et clairement au premier moteur universel, de qui vient egalement la suite et l'accord des perceptions.”

[101] Leibniz, *Eclaircissement des difficultés que Monsieur Bayle a trouvées dans le systeme nouveau de l'union de l'ame et du corps*, in ders., *Philosophische Schriften*, 265. – Das französische Original lautet (ebd., 264): “Je n'ay comparé l'ame avec une pendule qu'à l'egard de l'exactitude reglée, qui n'est même qu'imparfaite dans les ouvrages de Dieu; et on peut dire que l'ame est un Automate immateriel des plus justes.”

[102] Vgl. Leibniz, *Eclaircissement des difficultés*, 254-255: “Or c'est selon moy la nature de la substance créée, de changer continuellement suivant un certain ordre, qui la conduit *spontanément* [...] par tous les estats qui luy arriveront, de telle sorte que celuy qui voit tout, voit dans son estat present tous ses estats passés et à venir. Et cette

folgenden Abschnitt, die uniforme Aktivität der Seele darin, „beständig ein und demselben Gesetz der Ordnung oder des Fortgangs [...] zu folgen, wie in einer gewissen Zahlenreihe oder –folge [...]" ("suivre perpetuellement une même loy d'ordre ou de continuation, comme dans un certain rang ou suite de nombres [...].")[103]. Implizit wird durch den Vergleich der Seele mit einer *bestimmten* Reihe oder Folge von Zahlen auch die Leistungsfähigkeit der Uhr-Metapher erweitert. Die Pendeluhr vermag durch ihre Tätigkeit der exakten Sequenzierung der Zählstunden nun nicht nur die Präzision der Ordnung und Folge der Perzeptionen der Seele schlechthin zu repräsentieren, sondern auch, unter der Bedingung, dass ein Uhrwerk auf unzählig verschiedene Weisen mit unterschiedlichen Zählsequenzen eingerichtet werden kann, die individuelle Gesetzmäßigkeit einer Individualseele, der eine bestimmte, singuläre und einem spezifischen Gesetz der Ordnung und Abfolge gehorchende Zahlenreihe entspricht. Somit ist es die konzeptuelle, zweckgerichtete Kombination aus Mechanik, Arithmetik und qualitativer sowie quantitativer Disposition der Teile, die aus der Uhr ein technomorphes Sinnbild für den vollständigen Begriff, die *notio completa*, der individuellen Substanz macht, in der Gott als Wurzel (*radix*) der Zeit deren vergangene, gegenwärtige und zukünftige Prädikate eingefaltet hat, damit diese durch die raumzeitkonstruierende Tätigkeit der perzeptiven Substanz sukzessive entfaltet werden.[104] Die

loy de l'ordre qui fait l'individualité de chaque substance particuliere, a un rapport exact à ce qui arrive dans toute autre substance, et dans l'univers tout entier." ("Nun besteht nach meiner Meinung die Natur der geschaffenen Substanz darin, sich kontinuierlich zu verändern, einer gewissen Ordnung entsprechend, die sie *spontan* [...] durch alle Zustände führt, die ihr zustoßen werden, so daß wer alles sieht, in ihrem gegenwärtigen Zustand alle ihre vergangenen und künftigen Zustände erblickt. Und dieses Ordnungsgesetz, das die Individualität jeder besonderen Substanz bewirkt, steht in einer genauen Beziehung zu dem, was in jeder anderen Substanz und im ganzen Universum geschieht."

103) Leibniz, *Eclaircissement des difficultés*, 264-265. Vgl. dazu a. Hanns-Peter Neumann, „Atome, Sonnenstäubchen, Monaden. Zum Pythagoreismus im 17. und 18. Jahrhundert," in *Aufklärung und Esoterik. Rezeption – Integration – Konfrontation*, Hg. Monika Neugebauer-Wölk (Tübingen, 2008), 205-282, insbes. 244-252.

104) Vgl. u. a. Leibniz, *Principia logica-metaphysica*, AA VI 4B , 1645: "Notio completa seu perfecta substantiae singularis involvit omnia ejus praedicata praeterita praesentia ac futura."

individuelle Substanz ist ihrem Wesen nach also sukzessive Explikation des ewig Gleichzeitigen ihrer individuellen Natur, so wie die Uhr sukzessive Expression der kreisförmigen Ordnung der natürlichen Welteinrichtung ist. Das Horologium kann dadurch zwar auch als Transfermodell fungieren, das zwischen Körper und Seele, zwischen Welt und Gott vermittelt. Anders als bei Cusanus und Ficino aber, bei denen die Seele regulierender Agent des Körpers ist, anders auch als bei Descartes, bei dem die Seele zumindest auf den menschlichen Körper einen indirekten, willkürliche Bewegungen determinierenden Einfluss ausübt, hat bei Leibniz die metaphorische Transferleistung des Uhr-Modells aufgrund der radikalen Separation von Seele und Körper keinen *realen* Bezug mehr, sondern nur noch einen idealen. Das horologische Transfermodell ist zu einer ausschließlich ideellen Repräsentation und imaginären Metapher geworden, die allenfalls die innergöttliche Relation zwischen immateriellen Substanzen und körperlichen Phänomenen darzustellen vermag oder, in dualer Zersplitterung, die parallele Existenz zweier Welten demonstriert, zu deren übereinstimmender Justierung der Schöpfergott eine dritte Uhr appliziert hat: das perfekt normierende Präzisions- und Messinstrument seines mathematischen Verstandes.

Ungeachtet seines strikten Leib-Seele-Dualismus' erklärt Leibniz die Seelen bzw. Seelenmonaden nun aber auch zu Lebensprinzipien (*principes de la vie*) organischer Körper. Diese Lebensprinzipien unterscheiden sich aufgrund der Lehre von der präetablierten Harmonie naturgemäß von den plastischen, auf die Körper einwirkenden immateriellen Naturen, wie sie der Cambridger Platoniker Ralph Cudworth konzipiert hat. Da aber bei Leibniz ein unmittelbares kausales Verhältnis von Seele und Körper ausgeschlossen ist, stellt sich die Frage nach dem Verhältnis von Seele, Leben und Körper wie von selbst. Die beiden Uhren, die Seelenmaschine und die Körpermaschine, sind, wie bereits oben ausgeführt, zwei völlig voneinander unabhängige autonome Systeme, die nur durch einen göttlichen Akt von Anbeginn der Welt an parallel geschaltet wurden und parallel geschaltet bleiben. Leben als Signum der Differenz zwischen Materiemasse und organisierten Körpern bedeutet dann aber nichts anderes als die bloße unmittelbare Anwesenheit von formalen, finalursächlich geregelten Seelenautomaten in wirkursächlich regulierten Körpermaschinen, eine Anwesenheit freilich, die keinen konkreten gemeinsamen Berührungspunkt aufweist. Vielmehr ist das *interne* Verhältnis von Seelen- und Körpermaschine ein idealistisches: die Seelenmaschine *erzeugt* die Körpermaschine perzeptiv. Letztere ist somit nicht ein von außen erregtes Wahrnehmungsbild, auch wenn sie uns als solches *erscheinen* mag, sondern ein von innen generiertes, aus dem Inneren der perzeptiven Seele heraus konstruiertes Bild des eigenen Körpers. Die autonome Körpermaschine ist somit für die Seele mit dem perzeptiv konstruierten Bild dieser Maschine identisch, die die Seele von dem ihr von Gott anfänglich zugesprochenen Körper hervorbringt. Da aber allein und ausschließlich Gott die extramentale Existenz der Körperwelt garantiert und 'wahrnehmen' kann, ist die Entsprechung von perzeptivem Körper*bild* und dem tatsächlich extramental existierendem Körper zwar theologisch abgesichert. Weil Körper und Seele jedoch zwei völlig autonome, voneinander unabhängig existierende Systeme darstellen, die nicht aufeinander einzuwirken vermögen, kann dasjenige, was *wir* als körperlich wahrnehmen, nur ein *immanentes* Produkt der autarken perzeptiven Tätigkeit und Strebung der Seele sein. Körper müssen aus der Perspektive des immanenten Leib-Seele-Verhältnisses – aber eben nicht aus der Perspektive Gottes, die wir höchstens theoretisch anzunähern, jedoch nicht praktisch-empirisch einzunehmen vermögen – folglich immer Produkte der perzeptiven Tätigkeit der Seele sein. Sie lassen sich grob in zwei Klassen teilen: 1. leblose wirkursächlich nach den Regeln der Mechanik funktionierende Masseaggregate, in denen simple, immaterielle Einheiten mit einem bloß geringfügigem perzeptivem Vermögen anwesend sind; diese Masseaggregate erscheinen uns mitunter so, wie diese elementaren Monaden sich ihre Körpermasse selbst perzeptiv-konstruierend vorstellen mögen; und sie erscheinen uns überhaupt nur deswegen, weil Gott sie in uns ebenfalls schon immer *bildhaft* angelegt hat, d. h. weil unser eigener Körper eine graduelle Höherentwicklung eines solchen Masseaggregats darstellt; und 2. organische Körper, die wirk- *und* finalursächlich organisierte lebendige psychophysische Systeme vorstellen.

An der Schnittstelle von Mechanismus und Organismus wird so die perzeptive Konstruiertheit des Körpers durch die Seele besonders sinnfällig. Dort nämlich, wo eine *Seele* oder *Seelenmonade* anwesend ist, *scheint* der mechanische Körper in einer zweckursächlichen

Organisation seiner Teile, die intentional auf das Ganze der Leib-Seele-Einheit ausgerichtet ist, aufzugehen. Das perzeptive Körper*bild* des organischen Körpers, in dem eine Seele anwesend ist, so meine These, stellt demnach die Projektionsfläche dar, auf welche die individuelle Seele die jeweils nur ihr zukommende singuläre Finalursächlichkeit immer schon und stetig fortlaufend projiziert. Die Gesetzmäßigkeit der Seele, die in einer bestimmten Abfolge (*suite*) und Serie von Strebungen und Perzeptionen besteht, ist somit das regulative Ordnungsprinzip ihres *organischen* Körper*bildes*. Erst dadurch aber wird im Falle der Seele aus dem autonomen mechanischen Körper ein lebendiger, organischer Körper perzeptiv ge*bildet*, der schließlich auch nur als perzipiertes Bild repräsentiert wird, das sich deutlich vom Bild einer reinen, mechanisch geregelten Materiemasse unterscheidet. Als ein solches organisches Körperbild reaktualisiert sich das physische Erscheinungsbild einer bestimmten Seelenmonade auch in anderen Seelenmonaden, in denen es bereits durch den präetablierenden Akt Gottes implementiert ist und dadurch perzipierbar wird. Es ist somit das singuläre Reihengesetz einer Seele, das zum zweckursächlichen Muster ihres physischen Organismus wird. Wo also eine Seele ist, so lässt sich Leibniz' Auffassung pointiert zusammenfassen, dort perzipieren wir auch einen lebendigen Körper; wo wir einen lebendigen Körper perzipieren, dort muss auch eine Seele sein. Oder anders: Genauso wie eine Seele in einem Körper anwesend sein muss, damit *Leben* und *Lebendigkeit* entstehen und perzipiert werden können, muss der lebendige Körper das Körper*bild* selbst sein, das in der Seele anwesend ist, indem es durch deren perzeptiv-strebende Tätigkeit immer wieder generierend-regenierend repliziert und aktualisiert wird. Letzteres ist die Perspektive des monadologischen Phänomenalismus, ersteres diejenige der prästabilierten Harmonie, derzufolge Gott sowohl die Prozesse, Abfolgen und Verhältnisse der Monaden untereinander, der Körper untereinander und der Monaden und Körper zu- und miteinander eingerichtet und abgestimmt hat. So unabhängig aber die beiden Uhren der seelenmonadischen Maschine und der Körpermaschine hinsichtlich ihrer 'Architektur' voneinander auch sein mögen, so sind sie zur gleichen Zeit doch stark ineinander verschränkt und miteinander vernetzt. Schließlich treten die beiden Uhren immer, freilich in je unterschiedlichen Graden von der einfachen Monade bis zum komplexen Bewusstsein, vom Materieaggregat über den organischen Körper bis zum Geistkörper, miteinander und keineswegs unabhängig voneinander auf.

## 7. „Ich will es wieder durch das Geheimniss von der Uhr erläutern": Die multiple Funktion der Uhr-Metapher in der *Deutschen Metaphysik* von Christian Wolff

Bei Christian Wolff findet sich eine breit gefächerte, nahezu inflationäre Verwendung des Horologium als veranschaulichendes Exempel und als Metapher.[105] Für Wolff stellt die Uhr den Prototyp der zweckmäßig eingerichteten Maschine schlechthin vor. Sie ist dadurch im besonderen Maße geeignet, als Explikationsmodell für die Konstitution und Konzeption der Weltmaschine sowie als physikotheologisches Argument zu dienen. Dabei knüpft Wolff an Leibniz' Prinzip der Abfolge von perzeptiven und körperlichen Zuständen an, ohne Leibniz freilich bei Namen zu nennen und ohne dessen Vorstellung von der Seele als eines formalen Automaten zu übernehmen.

Bemerkenswert aber ist vor allem die Verbindung, die Wolff zwischen Welt, Wahrheit und Maschine zieht: „[...] so ist eben deswegen in der Welt Wahrheit, weil sie eine Maschine ist."[106] Meines Wissens nach ist dies das erste Mal, dass die Begriffe *Welt*, *Wahrheit* und

---

105) Geht es Wolff zum Beispiel um die Explikation des Begriffes der Ähnlichkeit, so führt er dem Leser zur Veranschaulichung zwei nahezu identische Uhren vor Augen. Vgl. Christian Wolff, *Vernünfftige Gedancken von Gott, der Welt und der Seele des Menschen, auch allen Dingen überhaupt* (= Deutsche Metaphysik), in ders., *Gesammelte Werke*, I. Abt.: Deutsche Schriften, Bd. 2.1 (Hildesheim/Zürich/New York, 2003), § 20. – Wie stark Wolffs Interesse an der horologischen Technologie gewesen ist, lässt sich auch daran ablesen, dass er veranlasst hat, Jacques Alexandres *Traité general des Horloges* ins Deutsche übersetzen zu lassen. Vgl. *Jacob Alexanders [...] Ausführliche Abhandlung Von Den Uhren überhaupt. Mit vielen Kupfern, Aus dem Frantzösischen in das Teutsche übersetzet [...] Von D. Christian Philipp Berger [...]* (Lemgo, 1738). Berger kommt, wie er in seiner Vorrede vermerkt, mit der Übersetzung einem ausdrücklichen, brieflich geäußerten Wunsch Wolffs nach, vgl. ebd., Vorrede des Übersetzers, 3v.

106) Christian Wolff, *Vernünfftige Gedancken von Gott, der Welt und der Seele des Menschen, auch allen Dingen überhaupt* (= Deutsche Metaphysik), in ders., *Gesammelte Werke*, I. Abt.: Deutsche Schriften, Bd. 2.2 (Hildesheim/Zürich/New York, 2003), § 559.

*Maschine* so unmittelbar zueinander in Beziehung gesetzt werden.[107] Wie ist diese Beziehung von Maschine, Wahrheit und Welt bei Wolff nun aber zu verstehen?

Zunächst rekurriert Wolff auf die Cartesische Begründung der Funktionsweise eines Körpers durch Disposition und Qualität seiner Teile, indem er das Wesen zusammengesetzter Dinge in der Art ihrer Zusammensetzung und in der Beschaffenheit ihrer Teile bestehen lässt.[108] Um die Vollkommenheit eines zusammengesetzten Dings beurteilen zu können, bedarf es – und hier unterscheidet sich Wolff von Descartes, indem er von vornherein *natürlich* gegebene Zweckursachen annimmt – der Einsicht in die Relation von funktionalem Zweck, Disposition und Konstitution der Teile eines zusammengesetzten Dings, die dessen final- und wirkursächliche Ordnung ausmachen. Demnach kann, so Wolff, die Ordnung einer Uhr auf unterschiedliche Weise eingerichtet sein, so dass diverse Vollkommenheitsgrade in deren Funktion der Zeitindikation erreichbar sind. Je detaillierter und präziser die Zeitindikation einer Uhr (von der rein stundenindizierenden bis hin zur sekundenanzeigenden Uhr), desto größer ist ihre Vollkommenheit und Ordnung in der Zusammensetzung, materialen Beschaffenheit und feinmechanischen Verarbeitung ihrer Teile.[109] In den §§ 553 bis 578 seiner *Deutschen Metaphysik*, in denen Wolff immer wieder auf die Uhr als Beispiel und Metapher zurückkommt, transferiert Wolff schließlich das Uhr-Modell auf die *machina mundi* bzw. auf das Konzept unzähliger möglicher Welten, deren Differenz auf die jeweilige Disposition und Konstitution ihrer Teile zurückzuführen ist. Die *reale* Welt ist die Folge der Realisierung einer *bestimmten* Welt aus vielen möglichen Welten, von der angenommen werden kann, dass sie, weil sie von Gott ins Sein gesetzt wurde, die beste resp. vollkommenste aller möglichen Welten ist.[110] Der Kritik an der *mundus optimus*-Theorie, Gott hätte

---

107) Selbstverständlich findet sich z. B. auch bei Ficino ein implizites Verhältnis von *machina mundi* und Wahrheit, aber keineswegs in einer solchen expliziten Formulierung wie bei Wolff.

108) Vgl. Wolff, *Deutsche Metaphysik*, § 9, § 155.

109) Vgl. Wolff, *Deutsche Metaphysik*, §§ 153-156.

110) Der Vergleich der vom Uhrmacher konzipierten, zum Zweck der Zeitindikation optimal eingerichteten Uhr mit der vom Schöpfer bestrealisierten Welt findet sich vielfach in Wolffs Werken, vgl. z. B. Christian Wolff, *Theologia Naturalis Pars II*, in ders.,

eine bessere Welt ohne die Existenz sündiger Menschen schaffen können, begegnet Wolff später ebenfalls mit dem veranschaulichenden Exempel des Horologium und des Uhrmachers.[111]

Vorerst und vornehmlich geht es Wolff aber um die Singularität dieser einen aktuell realisierten Welt, in der wir leben und deren Bewegungsablauf aus der einzigartigen Anordnung und Beschaffenheit ihrer Teile resultiert. Jede noch so geringe, von außen bewirkte willkürliche Änderung würde die Bewegungsabfolge in der Welt und die davon abhängigen Ereignissequenzen genauso ändern, wie eine Uhr anders laufen würde, entfernte man aus ihr „einen Theil, der zu seiner [*ihrer*] Bewegung gehöret […]."[112] Eine solche Veränderung hätte eine ganz andere Welt resp. Uhr und eine ganz andere Serie von Bewegungen zur Folge.[113] Eine *bestimmte* Uhr und eine *bestimmte* Welt sind demnach in sich geschlossene Systeme, die aufgrund der Einzigartigkeit ihrer zusammenwirkenden und zusammenstimmenden Komponenten mit einer besonderen, berechenbaren Sukzession von Bewegungen,

---

*Gesammelte Schriften*, II. Abt.: Lateinische Schriften, Bd. 8 (Hildesheim/New York, 1981), § 376: "Eodem modo artifex peritus constructurus horologium, non aliter idem construere potest, quam possibile per ideam in intellectu. Et dum construit optimum, quod nempe tempus omnium exactissime metitur; nec aliter ipsum construere vult. Quemadmodum nihil hinc infertur in praejudicium artificis, quod aliter horologium construere non potuerit; ita nec quicquam, quod Deo indignum sit, inde colligitur, quod mundum hunc aliter facere non potuerit, quia optimum ob sapientiam, qua gaudet, summam creavit." – „Auf gleiche Weise kan ein erfahrner Künstler, der eine Uhr verfertigen will, dieselbige nicht anders machen, als es vermöge des Begriffes in seinem Verstande möglich ist. Und indem er die beste verfertiget, welche nemlich die Zeit aufs allergenaueste abmisset, so wird er auch dieselbige nicht anders verfertigen. Wie nun hieraus dem Künstler kein Nachtheil entstehet, daß er die Uhr nicht anders hat verfertigen können: also kan man auch daraus nicht abnehmen, welches GOtt unanständig wäre, weil er diese Welt nicht anders hat machen können, indem er sie um seiner höchsten Weisheit willen als die beste erschaffen hat." Übersetzung aus: Christian Wolff, *Natürliche Gottesgelahrtheit II.1*, in ders., *Gesammelte Werke*, I. Abt.: Deutsche Schriften, Bd. 23.4 (Hildesheim/Zürich/New York, 1995), § 376.

[111] Vgl. Wolff, *Deutsche Metaphysik*, § 1061.

[112] Ebenda, § 553.

[113] Wolff wiederholt die Tatsache, dass, entfernt oder ändert man auch nur das geringste Teil einer Uhr oder der Welt, sich daraus ein völlig anders geartetes System ergebe, vielfach an verschiedenen Stellen, vgl. u. a. Wolff, *Deutsche Metaphysik*, §§ 556, 566-567, 1061.

Zuständen und Ereignissen einhergehen. Letzteres nennt Wolff den raumzeitlichen *nexus rerum*: „Und solchergestalt sind die Dinge in der Welt der Zeit nach mit einander verknüpft, weil sie dem Raum nach mit einander verknüpft sind.“[114] Für diesen raumzeitlichen *nexus rerum* in der Welt, in dem das Räumliche das Zeitliche bedingt, ließ sich kein besseres Sinnbild finden als das der Uhr, deren Zahnräder *räumlich* ineinandergreifen, um die Bewegung der Zeiger und damit die Indikation der *zeitlichen* Folge zu ermöglichen. Das Uhrwerk selbst wird bei Wolff folgerichtig zum Gleichnis für die raumzeitliche Beschaffenheit der Welt: „Es verhält sich die Welt nicht anders als wie ein Uhrwerck.“[115] Das Horologium ist aber nicht nur Gleichnis. Es ist prototypische Maschine. Und da nach Wolff die Welt ebenfalls eine Maschine ist, so *ist* die Welt wie jedes zusammengesetzte Ding, worunter auch lebende Körper zu rechnen sind, eine – freilich natürliche – horologische Raum-Zeit-Maschine: „Man darf sich aber um so viel weniger befremden lassen, daß ich von einem Uhrwercke oder einer Maschine ein Gleichniß gebe. Denn die Welt ist gleichfals eine Maschine.“[116] Die Applikation des Begriffs *Maschine* sowohl auf den horologischen Automaten als auch auf die Welt begründet Wolff damit, dass beide, Welt und Uhr, zusammengesetzte Dinge sind – allerdings mit einer kleinen semantischen Nuance: Die Maschine ist, so Wolff, ein „Werck“, die Welt „ein zusammengesetztes Ding“.[117] Als Werk ist die Maschine ein *künstliches* Ding aus Menschenhand, als Kompositum ist die Weltmaschine ein *natürliches* Ding aus der Hand Gottes. Die Weltmaschine besteht aus Teilen, die für sich wiederum natürliche Maschinen sind, während die künstliche Maschine aus Teilen zusammengesetzt ist, die selbst keine Maschinen mehr sind. Die Naturmaschine selbst aber ist Bedingung für die mögliche Konstruktion jeder künstlichen Maschine, indem sie zum einen deren materiale Ursache stellt, zum andern die von Gott in ihr installierten Ordnungsformen (etwa die Planeten-

---

114) Ebenda, § 554. – Vgl. a. den Titel von Wolffs 1724 erschienenem Werk *De differentia nexus rerum sapientis et fatalis necessitatis, nec non systematis harmoniae praestabilitae et hypothesium Spinosae luculenta commentatio, in qua simul genuina Dei existentiam demonstrandi ratio expenditur et multa religionis capita illustrantur.*

115) Ebenda, § 556.

116) Ebenda, § 557.

117) Ebenda, § 557.

bewegung) vorgibt, an denen sich der *artifex*, der das in der Natur realisierte göttliche Konzept einsieht, orientieren kann, um ein *artificium*, wie z. B. das Horologium, zu entwerfen. Somit kann auch die künstliche Zusammensetzung eines *artificium* nichts leisten, was über die Gesetze der Natur hinausgeht. Hier, in dieser feinen Nuancierung, klingt auch bereits jenes von Wolff in späteren Paragraphen explizierte Partizipationsverhältnis zwischen menschlichem Erfindungsgeist und göttlichem Intellekt, zwischen Kunst und Natur an.[118]

Wolff weist nun den Zusammenhang von Wahrheit, Welt und Maschine am Begriff der Ordnung aus. Die maschinelle Struktur der Welt besitzt eine fixierte Ordnung, in der sämtliche Bewegungen, Veränderungen, Ereignisse und Zustände ineinander gegründet sind und auseinander hervorgehen. Alles hat seinen zureichenden Grund, warum es eher ist und geschieht als nicht ist und nicht geschieht. Jegliche vergangene, gegenwärtige und zukünfige Abfolge in den Zuständen der Welt ist wie diejenige der durch die Uhr indizierten Zeitmomente begründbar und erfüllt die Bedingung einer logischen Beweiskette nach dem Vorbild der euklidischen Geometrie. Die natürlichen und die künstlichen Dinge besitzen also eine mathematische und argumentationslogische Ordnung: „[...] so ist in der Welt und ihren Veränderungen eine Ordnung, nehmlich eben dergleichen Ordnung als man in einem ordentlichen Beweise und in den Schriften Euclidis antrift, folgends auch Wahrheit."[119] Ordnung aber besteht in der besonderen Disposition und Konstitution der Komponenten eines zusammengesetzten Dings zu einer selbstregulierten Sukzession von Bewegungen, Ereignissen und Zuständen. Sie macht das Wesen eines zusammengesetzten Dings aus, bestimmt aber auch jede argumentative Beweisführung: Die sukzessive, diskursive, nach logischen Regeln verfahrende Bewegung des Verstandes führt diesen von den Prämissen zur Konklusion und von da aus zu einem evidenten Nexus von bewiesenen Aussagen, der diese zu einem quasi organischen Ganzen ordnet. Die ordnende Kraft des Verstandes komponiert und koordiniert die diversen bewiesenen Aussagen zu einem in sich geschlossenen Ordnungssystem und begründet damit deren Wahrheit. Da die Maschine *per*

118) Vgl. ebenda, §§ 995-996.

119) Ebenda, § 558.

*definitionem* die genannten Ordnungskriterien der Komposition und Koordination der disponierten Teile zu einem funktionalen Ganzen erfüllt, die Welt aber eine in sich geordnete und wohlorganisierte Maschine ist, so ist „in der Welt Wahrheit, weil sie eine Maschine ist.“[120] Dinge sind aufgrund ihrer Zusammensetzung folglich geordnete Maschinen. Also *sind* und haben sie „Wahrheit“. Es sind nun aber gerade die Dinge und deren Zustände und Sequenzen, auf denen jegliche logische und mathematische Wahrheit beruht. Die Wissenschaften der Mathematik und Logik fungieren demzufolge als deskriptive Explikations- und Erschließungsmodelle des ‘maschinellen’ *nexus rerum* in der Welt. Sie stellen den Versuch dar, den in sich rational strukturierten horologischen Welt-Automaten, seine autonome Geschlossenheit und Selbstregulation, in einem ebenso geschlossenen logisch-mathematischen System abzubilden und zu erschließen. Für eine logische Hermeneutik, wie sie von Wolff als wissenschaftliches Instrument entwickelt worden ist, bedeutet das, dass die Nominaldefinitionen, also die arbiträren Wort- und Begriffsbildungen und -erklärungen, in Realdefinitionen, d. h. in die Bestimmungen der Dinge in ihrem sachlichen *nexus*, aufgelöst resp. die Worte auf die Dinge zurückgeführt werden müssen, um durch die Worte „Wahrheit“ abbilden zu können.

Weil die Welt eine horologische Maschine ist, hängen auch alle zeitlichen Modi in ihr notwendig zusammen:

> Weil der gegenwärtige Zustand der Welt im vorhergehenden und der zukünftige in dem gegenwärtigen gegründet ist; so erhalten dadurch die Begebenheiten in der Welt ihre Gewißheit. Und solchergestalt sind dadurch, daß die Welt eine Maschine ist, alle Begebenheiten darinnen gewiß gemacht.[121]

Die Reihe der Zustände und Begebenheiten nennt Wolff an anderer Stelle auch „unzertrennete Reihe“: „Ja, das Exempel der Uhr zeiget, man habe auch auf alle zukünftigen Zeiten zu sehen, die in einer unzertrenneten Reihe auf die gegenwärtige folgen."[122] Es ist unschwer zu erkennen, dass Wolff mit der horologischen Metapher die *notio completa*

120) Ebenda, § 559.

121) Ebenda, § 561.

122) Ebenda, § 565.

der Welt versinnbildlichen möchte, deren vollständigen Begriff, in dem die singuläre *Reihe* und *Abfolge* an Zuständen, Ereignissen, Dingen dieser einen besonderen Welt – im Unterschied zu durchaus möglichen anderen denkbaren Welten und Reihen – eingefaltet ist. Durch den Schöpfungsakt wird aus der mentalen Konzeption des vollständigen Begriffs dieser einen besonderen Welt im Geist Gottes die sich prozessual entfaltende reale Existenz dieser einen besonderen Welt. Während die reguläre Reihe dieser einen realen Welt, in der wir leben, gewiss und wahr ist, ist sie als ideale, konzeptuelle Welt, als nur eine unter vielen möglichen Welten hingegen keineswegs absolut notwendig. Ihre Notwendigkeit ist nur eine singuläre, relative. Zu jedem weltlichen *de facto*-Ereignis wären nämlich andere, alternative Ereignisse denkbar, die jedes für sich eine besondere, singuläre, nach einem eigenen seriellen Gesetz regulierte Welt voraussetzen und repräsentieren würden. Auch hier nutzt Wolff wiederum das sinnbildliche Potenzial der Uhr-Metapher, um das bereits von Leibniz herausgestellte Prinzip der Abfolge von Zuständen in seinen Implikationen und Konsequenzen zu verdeutlichen: „Man feile von dem kleinesten Zahne eines Rades was ab, so wird nach diesem die Uhr beständig anders gehen, als sie würde gegangen seyn, wenn man nichts weggefeilet hätte, und sie ist auch in der That nicht mehr dieselbe Uhr."[123] Ein anderer denkbarer Zustand als der tatsächlich eingetretene verlangt also eine andere innere Disposition und Konstitution der Komponenten des jeweiligen Kompositums und bedingt somit auch eine andere Identität. Diese andere Identität wiederum beruht auf dem Gesetz einer zur aktuellen Ereignisserie *dieser* Welt alternativen „unzertrenneten Reihe" einer anderen Weltkonzeption.

Nun gehören nach Wolff Raum und Zeit zu den idealen, imaginären Begriffen des menschlichen Intellekts; sie sind also Resultat des spezifischen Erkenntnismodus menschlichen Denkens, nicht aber extramental existierende Entitäten. Das heißt aber, dass die horologische Raum-Zeit-Maschine *Welt* in der Sukzessivität ihrer Ereignisse und Zustände ein Produkt des begrenzten, bloß zur Diskursivität befähigten Erkenntnisapparats des Menschen ist.

---

123) Ebenda, § 568.

Der menschliche Verstand selbst aber steht wiederum in einem diskursiven Annäherungs- und Partizipationsverhältnis zum göttlichen, alles auf einmal und zugleich übersehenden und vorherwissenden Intellekt, der die *notiones completae* aller Dinge enthält.[124] Prägnant formuliert Wolff das Modell der diskursiv und progressiv annähernden Entfaltung des göttlichen Intellekts durch menschliches Denken, Tun und Invenieren in einem Abschnitt zu Huygens Erfindung der astronomischen Pendeluhr:

> Die Planeten-Uhr, welche Hugenius erdacht, ist nicht erst möglich worden, da er derselben nachgedacht, sondern schon von Ewigkeit her mit in dem göttlichen Verstande gewesen, der sie wie alle übrige Wahrheiten hervorgebracht. Und wenn wir an der Zusammensetzung derselben, oder einer andern Sache etwas ändern; so ändern wir nicht das Wesen des Dinges, obgleich dasselbe in der Art der Zusammensetzung bestehet; sondern weil wir nach und nach dencken müssen, so kommen wir von dem Wesen einer Planeten-Uhr, oder überhaupt eines Dinges auf das Wesen eines anderen Dinges, das von ihm bloß darinnen unterschieden ist, was geändert wird. Wüste ein Mechanicus, der eine Planeten-Uhr machen will, alle Planeten-Uhren, die möglich sind, und könnte alle insgesamt sich auf einmahl vorstellen; so dürfte er nur daraus diejenige erwehlen, die zu seiner Absicht sich am meisten schickte. Und so muß man sich die Sache auch bey GOtt vorstellen [...]. Und hieraus erkennet man, daß GOtt allein der einige und wahre Erfinder ist, als dessen Verstand alles, was möglich ist, hervorbringet: hingegen die Menschen nur von den unendlichen Erfindungen GOTTES eine und die andere zu schauen bekommen, indem sie ihre Gedancken darnach richten.[125]

Bei der multifunktionalen Anwendung der Uhr-Metapher in der *Deutschen Metaphysik* – als kosmologisches Gleichnis; als Beispiel für die Konzeptualisierung aller Dinge im Geist Gottes, an dem der

124) Vgl. ebenda, § 1029: „Z. E. Der Erfinder der Uhren hat zu seiner Absicht gehabt die Erkäntniß der Zeit. Unterdessen da die Uhren sind gemacht worden, hat man sie auch gebraucht einen Staat damit zu machen. Und da die Erkäntniß der Zeit in vielen Fällen nöthig befunden worden; so haben die Uhren vielfältigen Nutzen bekommen, daran der Erfinder der Uhren gar nicht gedacht hat. Und dannenhero ist der Nutzen eines Dinges in Ansehung der Menschen von seiner Absicht unterschieden. Hingegen kan dieses bey GOtt nicht statt finden. Er übersiehet alles und weiß vorher, was daraus aus allen möglichen Umständen erfolgen kan. Da er nun auch alles zu erhalten gedacht; so ist aller Nutzen der Dinge eine göttliche Absicht, ohne welche er sie nicht würde erwählet haben."

125) Ebenda, §§ 995-996.

invenierende Verstand des Menschen partizipiert; als Zeugnis für den Raum und Zeit erst diskursiv konstruierenden menschlichen Intellekt und als naturtheologisches resp. physikotheologisches Argument –[126] überrascht indes, dass Wolff, anders als Leibniz, das horologische Vergleichsmodell nicht heranzieht, um die Seele selbst zu repräsentieren. Und doch wird sie insofern implizit mitgedacht, als Raum und Zeit von Wolff als Konstrukte der perzeptiven Tätigkeit der Seele vorgestellt werden. Unter die perzeptive Aktivität der Seele müssen nun aber auch das Auseinander, Miteinander und Ineinander der Komponenten eines Kompositums, die Maschine als Ordnungssystem, für die bei Wolff die prototypische horologische *machina* steht, subsumiert werden. Maschine-Sein und Welt-Sein dependieren folglich von der Beschaffenheit des Perzeptionsvermögens der Seele: Die von Gott *de actu* eingerichtete Schöpfung samt der Fülle der vollständigen Begriffe (*notiones completae*) entfaltet sich, folgt man der Argumentation Wolffs, demnach erst im Durchgang durch die repräsentative Kraft der perzeptiven Fähigkeit der Seele (*facultas percipiendi animae*) in und als die uns wahrnehmbare und von uns reflektierbare raumzeitliche Weltmaschine.[127]

Da ein zusammengesetztes Ding resp. eine Maschine aus Teilen bestehen muss, die Seele hingegen als immaterielle Unität keine Teile hat, so ist es nur konsequent, dass für Wolff die perzeptive Seele zwar als maschinekonstituierendes und -konstruierendes Wesen, nicht aber selbst als Maschine gelten darf. So wie Gott erste Wirkursache (*causa prima efficiens*) oder U(h)r-Macher der Schöpfung ist, fungiert die perzeptive Seele als *causa efficiens* der Weltrepräsentation in Raum und Zeit. Die horologische Weltmaschine tritt folglich erst im Moment des Perzipiertwerdens, des aktiven Repräsentiertwerdens den Weg vom

[126] Wolff nutzt die Uhr in seinen naturphilosophischen, insbesondere physiologischen Werken, die auf der in der *Deutschen Metaphysik* gelegten ontologischen Basis aufbauen, auch als Metapher für organische Körper; vgl. dazu Tobias Cheung, „Ontologie und Physiologie bei Christian Wolff (1679-1754)," *Verhandlungen zur Geschichte und Theorie der Biologie*, 10 (2003), 263-281, insbes. 275-279 („Von Uhren, Gebäuden und 'vielfältigen Maschinen'").

[127] Im Grunde klingt schon bei Leibniz und Wolff das Kantsche Konzept des Dings resp. der Welt an sich an: Denn da die Welt, wie wir sie wahrnehmen, nur deswegen in Raum und Zeit besteht, weil Raum und Zeit Konstruktionen der menschlichen Erkenntnisaktivität sind, so ist die Weltmaschine nichts anderes als ein limitiertes Annäherungsmodell an das raumzeitunabhängige Schöpfungskonzept im Geist Gottes.

Möglichen ins Wirkliche an. Wirklich aber, *real*, wird sie erst durch ihre in den Perzipienten göttlich angelegte und von diesen dann perzeptiv reaktualisierte und realisierte prozessuale Ordnung, deren intersubjektive Übereinstimmung ihre transsubjektive göttliche Wahrheit ausmacht. Die rationale Logik dieser Ordnung korrespondiert dann mit der Struktur der horologischen Weltmaschine, wie sie im göttlichen Geist, an den sich der Mensch denkend annähert, konzipiert worden ist.

## 8. Ausblick und Resümee: La Mettrie, Ploucquet und Diderot

Julien Offray de La Mettrie hat in seiner aufsehenerregenden Ende 1747 publizierten, auf 1748 vordatierten Schrift *L'Homme Machine* die Uhr nicht nur als Modell des menschlichen Körpers verwendet. Er hat auch die Seele selbst zu einem materiellen Bestandteil des horologischen Körperautomaten erklärt: „Die Seele ist nur ein Bewegungsprinzip oder ein empfindungsfähiger materieller Teil des Gehirns, den man, ohne einen Irrtum zu fürchten, als eine Hauptfeder der ganzen Maschine ansehen kann."[128]

Um die Frage beantworten zu können, was das eigentlich Neue und Aufsehenerregende dieser Definition La Mettries war, bedarf es eines kurzen abschließenden Rückblicks auf die schon eingehend beschriebenen Interpretationen der Uhr als Metapher und Begriff von der Renaissance bis zur Leibniz-Wolffschen Philosophie. Die bisher skizzierten Funktionen des horologischen Vergleichsmodells, wie sie bei Cusanus, Ficino, Descartes, Leibniz und Wolff sichtbar wurden, können so als komparative Folie dienen, um den Bruch zu beschreiben, den Julien Offray de La Mettries Transformation des Seelenbegriffs und des Leib-Seele-Verhältnisses markiert:

---

128) Julien Offray de La Mettrie, *L'Homme Machine* (Leiden, 1748), 84: "[...] l'âme n'est qu'un principe de mouvement, ou une partie matérielle sensible du cerveau qu'on peut, sans craindre l'erreur, regarder comme un ressort principal de toute la machine [...]." – Diese Ausgabe ist von der Berlin-Brandenburgischen Akademie der Wissenschaften digitalisiert worden, siehe online unter: http://bibliothek.bbaw.de/quellendigital/lamettrie/metr_homm_fr_1748/index.html

(1) Die Momente der Selbstregulation und der zirkulären Selbstbewegung des Horologium, in dessen zeitindizierendem Lauf sich das technische Genie des Uhrmachers simulativ fortsetzt, stehen bei Cusanus und Ficino für das den Körper verlebendigend-bewegende und beherrschende Vermögen der ihrerseits zur Selbstbewegung befähigten immateriellen Seele. Da die Seele sowohl am göttlichen Geist als auch an der 'weltlichen' Materie partizipiert, vermag die Uhr zudem als Metapher für das einfaltend-entfaltende Verhältnis von Gott und Welt zu dienen, zwischen denen die Seele als dynamische und Leben-generierende Mittlerin positioniert ist.

(2) Aber auch Descartes sieht in der Uhr ein Sinnbild für Bewegung und Selbstbewegung. Die Selbstbewegungskräfte sind bei Descartes jedoch Ausdruck der mechanischen Gesetzmäßigkeiten der Materie selbst. Das Horologium versinnbildlicht folglich ausschließlich die selbstregulativen Mechanismen der Körpermaschine. Die Seele kann somit mit und seit Descartes nicht mehr länger und ohne Weiteres zum unbezweifelten Bewegungsprinzip des Körpers deklariert werden. Nur mittelbar bleibt sie an den Bewegungen und Handlungen beteiligt, indem sie die willkürlichen Bewegungen des Menschen rational determiniert, ohne diese indes ausführen zu können, während die ihr traditionell zugeschriebene vegetative und sensitive Funktion der Regulation unwillkürlicher Bewegungen durch die 'Automatik' des Körpers ersetzt wird.

(3) Leibniz führt beide Funktionen der horologischen Metapher, das Horologium als Seelenautomat und Körpermaschine, in seinem Zwei-Uhren-Modell zusammen, ohne sie aber in ihrem ursprünglichen Sinn zu übernehmen. Seelenuhr und Körperuhr stellen bei Leibniz voneinander unabhängige autonome Selbstbewegungs- und Selbstregulationsprinzipien vor: die eine ist finalursächlich, die andere wirkursächlich organisiert. Weder reguliert die Seele – und hier unterscheidet sich Leibniz deutlich von Cusanus und Ficino – den Körper, noch gesteht Leibniz eine wenn auch noch so schwache mittelbare Verbindung zwischen der *anima rationalis* und dem Körper zu, wie Descartes das getan hat (Animalgeister, Zirbeldrüse). Obwohl sich bei Leibniz keine reale, wirkursächliche Verschränkung der beiden Uhren findet, so doch, auf der Basis des monadologischen Phänomenalismus, eine ideale, in der die Körperuhr von der Seelenuhr als Körperbild perzeptiv *erzeugt* und

aus dem von Gott in ihr bereits angelegten Archiv an repräsentierbaren Perzeptionen *evoziert* wird.

(4) Wolff gebraucht die Uhr als Sinnbild für die Ordnung des Kompositums und der ineinandergreifenden Bewegung der zusammengesetzten Teile zu immer neuen und zugleich sich replizierenden Momentanzuständen des gesamten Weltsystems. Da die Seele kein Kompositum ist, kann auch die Uhr-Metapher, wie sie von Wolff interpretiert wird, die Seele nicht länger mit einbegreifen, wie das noch bei Leibniz möglich war. Die von Wolff ausdrücklich herausgestellte enge Verbindung des Ordnungs- und Maschinenbegriffs mit dem Begriff der Wahrheit zeigt aber, dass die Welt gerade in der Komposition und Koordination ihrer Teile durchgehend rational strukturiert ist. Verstandestätigkeit im Sinne der mathematisch-logischen Erschließung der Welt sowie die Welt selbst als horologisch-prozessuale Komposition der in ihr bereits existierenden und noch möglichen zukünftigen Dinge und Zustände konvergieren folglich im Begriff der Wahrheit. Da nun Raum und Zeit bei Wolff gleichermaßen Ordnungsprinzipien der Weltmaschine *und* Resultate perzeptiver Seelentätigkeit sind, ist die Ordnung der Welt eine perzeptive Konstruktion, die, wenn die diese Konstruktion beschreibenden Aussagen bewiesen sind, 'Wahrheit' hat, d. h. mit dem göttlichen Schöpfungskonzept, dem Begriff der Welt im Geist Gottes, korrespondiert.

Die unterschiedlichen Funktionalisierungen des horologischen Vergleichsmodells indizieren also, wie gezeigt, auch bestimmte Seelenbegriffe bei den zitierten Autoren. Es sind diese diversen Seelenkonzepte, von denen sich La Mettrie absetzt, indem er das Horologium als Metapher und Begriff reinterpretiert resp. umfunktionalisiert: Aus dem immateriellen Bewegungs- und Selbstbewegungsprinzip *Seele* bei Cusanus und Ficino; aus Descartes' unkörperlicher Seele, die nur mittelbar über die Animalgeister und nur beim Menschen für die *willkürlichen* Bewegungen des Körpers verantwortlich ist; aus dem autonomen und autarken formalen Seelen-Automaten bei Leibniz, der sein eigenes finalursächliches, ihn regulierendes Gesetz auf das von ihm perzeptiv erzeugte Bild des ihm über die göttliche präetablierte Harmonie parallel geschalteten Körpers projiziert; aus Wolffs erkenntnistheoretischer Affiliation von Wahrheit, Maschine und Welt, die aus der rational diskursiven Annäherungsbewegung des menschlichen Intellekts an den göttlichen Geist heraus allererst als horologische Maschine *konstruiert* wird – aus all diesen La Mettrie mehr oder weniger vertrauten, mit dem Uhr-Modell assoziierten und zur horologischen Metapher akkumulierten Leib-Seele-Auffassungen entsteht bei La Mettrie ein integratives materialistisch-monistisches Modell der materiellen körperrhythmisierenden, zeitlichen Taktgeberin und „Hauptfeder" ("ressort principal") *Seele*.

Der materialistisch-monistische Seelenbegriff La Mettries, der in deutlichem Gegensatz und in Konkurrenz zum monadologisch-monistischen Phänomenalismus bei Leibniz und Wolff steht, wird von La Mettrie nun wie folgt erläutert: Im ausdrücklichen Anschluss an Galens Verortung der *anima* im Gehirn bestimmt La Mettrie die sensitive Seele als neurophysiologisches impulssetzendes und – verarbeitendes Prinzip und materielle Taktgeberin des Körpers. Aus der Seele, die La Mettrie als horologische Hauptfeder des Körpers beschrieben hat, entspringt durch Emanation ("émanation") bzw. durch Kraftübertragung die „natürliche einem Pendel vergleichbare Oszillation [*eine Art Vibration*]" ("oscillation naturelle [...] semblable à celle d'une pendule") jeder Faser ("fibre") und jedes Faserelements ("Element fibreux") des Körpers, der, so La Mettrie, „nur eine Uhr ist" ("Le corps n'est qu'une horloge").[129] Noch in seinem 1745 in La Haye erstmals erschienenem, 1747 in zweiter Auflage ediertem Werk *L'Histoire Naturelle de l'Ame*, an dem er mit seiner Schrift *L'Homme Machine* anknüpft, hatte La Mettrie die sensitive Seele mit dem Schlag bzw. der Glocke oder dem tickenden Laut (*timbre*), nicht aber, wie in *L'Homme Machine*, mit der Hauptfeder einer Uhr (*montre*) verglichen: „Die Seele ist in gewisser Weise das im Körper, was die Glocke in der Uhr ist."[130] Jedoch ging La Mettrie in der *Histoire Naturelle de l'Ame* auch nur auf das rezeptiv-sensitive

---

[129] Ebenda, 84.

[130] Julien Offray de La Mettrie, *L'Histoire naturelle de l'Ame* (Paris, 1747), 47: "[...] l'Ame est en quelque sorte dans le corps, ce qu'est le timbre dans la montre." – Zum Sitz der Seele im Gehirn vgl. ebd., 43: "[...] c'est effectivement dans le cerveau, que l'Ame est affectée des sensations propres à l'Animal [...]." ("Es ist tatsächlich im Gehirn, dass die Seele von Empfindungen affiziert wird, die einem Lebewesen eigen sind."). Schon Galen hatte dies erkannt, so dass es, wie La Mettrie herausstellt, genüge, wenn man „immer wieder auf Galen zurückkäme wie zur Wahrheit selbst", ebd., 45: "Il en faudroit toûjours venir à Galien, comme à la vérité même."

Vermögen des „materiellen Prinzips" ("principe matériel") oder der „substanziellen Form" ("forme substantielle") des Körpers ein.[131] Demnach sind Wahrnehmung und Empfindung (*sensation*) der sensitiven Seele völlig von dem „Klöppel" ("marteau") abhängig, der durch die reguläre Bewegung der Federn und Räder des als Uhrwerk versinnbildlichten Körpers reguliert wird und schließlich den Klang bzw. die Glocke (*timbre*) – die „Seele" – des horologischen Körper-Automaten auslöst.[132] Von dieser rein passiven Vorstellung der Seele ist La Mettrie schließlich in *L'Homme Machine* dazu übergegangen, auch und vor allem die tätige Funktion der materiellen Seele zu perspektivieren und diese mit dem Hippokratischen *enormon*, dem im Körper aktiven dynamischen Lebensprinzip, zu identifizieren.[133] Auf diese Weise hebt La Mettrie den 'klassischen', vielfach variierten Seelenbegriff auf ("Die Seele ist nur ein leerer Begriff")[134] und macht aus der Seele – und hier klingt gewissermaßen die antike Lehre vom Seelenpneuma als Hauchseele an[135] – ein körperliches Nervenprinzip bzw. Nervenpneuma, das seinen Sitz im Gehirn am Ursprung der Nerven hat: „Dieses Prinzip

131) Ebenda, 40. – Es ist bezeichnend, dass La Mettrie den traditionellen Begriff der immateriellen substanziellen Form transformiert, indem er diese 'materialisiert'. Ich halte dies für eine kritische Antwort La Mettries sowohl auf den aristotelischen Substanzbegriff als auch auf den Leibnizschen Begriff der Monade.

132) Ebenda, 47-48: "Tout le corps de cette machine, les ressorts, les rouës ne sont que les instrumens, qui par leurs mouvemens concourent tous ensemble à la régularité de l'action du marteau sur le timbre, qui atend [sic!] pour ainsi dire, cette action, & ne fait que la recevoir: car lorsque le marteau ne frape pas le timbre, il est comme isolé de tout le corps."

133) *Enormon* leitet sich vom Altgriechischen Verb *hormao* ab, das *antreiben* bedeutet.

134) La Mettrie, *L'Homme Machine*, 71: "L'Ame n'est donc qu'un vain terme."

135) Vgl. dazu die immer noch unübertroffene Überblicksdarstellung von Franz Rüsche, *Das Seelenpneuma. Seine Entwicklung von der Hauchseele zur Geistseele* (Paderborn, 1933); vgl. außerdem Gérard Verbeke, *L'évolution de la doctrine du pneuma: du stoicisme à S. Augustin* (Paris, 1945); Marielene Putscher, *Pneuma, Spiritus, Geist. Vorstellungen vom Lebensantrieb in ihren geschichtlichen Wandlungen* (Wiesbaden, 1974). Zum Seelenpneuma im Mittel- und Neuplatonismus vgl. Werner Deuse, *Untersuchungen zur mittelplatonischen und neuplatonischen Seelenlehre* (Wiesbaden, 1983), v.a. 218 ff.

existiert und hat seinen Sitz im Gehirn am Ursprung der Nerven, mittels derer es seine Herrschaft auf den restlichen Körper ausübt."[136]

Mag der menschliche Körper nun zwar 'nur' eine über ein neurophysiologisches (Selbst)Bewegungsprinzip gesteuertes Uhrwerk sein, so steht für La Mettrie jedoch außer Frage, dass dieses Uhrwerk ein „ungeheures" ("immense") Kunstwerk vorstellt, das selbst, wenn einige seiner Teile versagen sollten, aus eigenem Antrieb, also selbstregulativ, weiter zu funktionieren vermag.[137] Der menschliche Körper ist, so La Mettrie, eine Maschine, „die ihre eigenen Federn aufzieht" ("qui monte elle même ses ressorts") und dadurch zum „lebenden Bild unaufhörlicher Bewegung" ("vivante image du mouvement perpetuel") wird.[138] Auch hier nutzt La Mettrie das tradierte Potenzial der Uhr-Metapher: War diese ursprünglich ein metaphysisches und theologisches Modell, das, wie bei Cusanus und Ficino, für die ideale Simulation der in der göttlichen Einheit gegründeten perfekten, seelengesteuerten permanenten Kreisbewegung der Himmelskörper und des Lebens schlechthin stand, oder, wie bei Descartes, ein physiologisches Modell, das die nicht seelisch, sondern mechanisch regulierte Selbstbewegung des Blutkreislaufs versinnbildlichte, so 'entmetaphysiert' La Mettrie einerseits die aus der Renaissance tradierte Uhr-Metaphorik und reduziert andererseits, in kritischer Distanzierung zu Descartes, sämtliche „Seelenvermögen" auf das Organisationsprinzip des Gehirns.[139] Die neurophysiologisch regulierte Körpermaschine *Mensch* nennt La Mettrie dann konsequent eine „wohl aufgeklärte Maschine" ("une Machine bien éclairée")[140], deren Denkvermögen und Bewusstsein das Resultat einer quasi hylozoistischen Evolution der Materie ist. Insofern verhält sich der Mensch zum Affen wie „die Planetenuhr von Huygens zu einer Uhr von Julien le Roy";[141] er ist aus demselben Stoff gemacht, aber graduell höher

136) La Mettrie, *L'Homme Machine*, 78: "Ce principe existe, & il a son siége dans le cerveau à l'origine des nerfs, par lesquels il exerce son empire sur tout le reste du corps."

137) Ebenda, 93.

138) Ebenda, 14.

139) Ebenda, 70: "[...] qu'elles [les facultés de l'Ame] ne sont visiblement que cette Organisation [du Cerveau] même [...]."

140) Ebenda, 70.

141) Ebenda, 92: "Il est au Singe, aux Animaux les plus spirituels, ce que le Pendule Planétaire de Huygens, est à une Montre de Julien le Roi."

entwickelt. Während die Leibnizianer, wie La Mettrie betont, die Materie eher spiritualisiert als die Seele materialisiert hätten, hat er selbst Seele und Geist materialisiert und die Seele zu einem körperlichen neurophysiologischen Bewegungsprinzip des horologischen Körperautomaten *Mensch* gemacht.[142]

Es ist reichlich bekannt, dass La Mettries atheistischer Materialismus seinerzeit breit diskutiert worden ist. Unter anderem hat auch der Leibnizianer Gottfried Ploucquet in mehreren Dissertationen auf La Mettrie reagiert.[143] Interessant ist im Zusammenhang mit dem Gebrauch der Uhr-Metapher vor allem eine kleine präformationstheoretische Schrift Ploucquets, die er auf der Grundlage eines vor der Berliner Akademie im August 1748 gehaltenen Vortrags 1749 in Stuttgart veröffentlicht hat: *De Corporum Organisatorum Generatione Disquisitio Philosophica*. Hierin argumentiert Ploucquet eingehend für ein immaterielles monadisches, bildreguliertes Organisationsprinzip des organischen Körpers, dessen Bewegungsablauf zwar mechanisch expliziert werden kann; die zweckgerichtete Koordination der Organe zu einem Ganzen aber und die Regenerationsfähigkeit lebender Organismen – besonders im Hinblick auf die gattungsspezifische Ähnlichkeit von Eltern und Nachkommen – sind, so Ploucquet, durch die *regulae motus* nicht hinreichend erklärbar. Das Explikationsdefizit des Mechanismus veranschaulicht Ploucquet am Beispiel der Uhr:

> Wahr ist freilich die Aussage: Wo es Organisation gibt, gelten die Regeln der Bewegung. Kehrst du aber diese Aussage um, dann irrst du dich. Es folgt nicht umgekehrt: Wo die Regeln der Bewegung gelten, dort gibt es Organisation. Die Bewegungsgesetze bringen es nicht mit sich, dass sie geordnete Wirkungen geordnet und konstant hervorbringen. Außerdem muss man auf die präexistente Koordination der Teile achten und darauf, dass diese in der Lage ist, eine solche Wirkung zu erzeugen. Eine Uhr bewegt sich gemäß den Regeln der Bewegung. Trenne aber die Räder, Gewichte etc. von ihr ab und sieh zu, ob diese durch die

142) Ebenda, 2: "Les Leibnitiens, avec leurs *Monades*, ont élevé une hypothèse inintelligible. Ils ont plûtot spiritualisé la Matière, que matérialisé l'Ame."

143) Zu Ploucquet und seiner Teilname an der Preisfrage der Berliner Akademie der Wissenschaften zu Berlin zur Monadologie, die auf das Jahr 1747 ausgeschrieben wurde, vgl. Hanns-Peter Neumann, „Zwischen Materialismus und Idealismus. Gottfried Ploucquet und die Monadologie," in *Der Monadenbegriff zwischen Spätrenaissance und Aufklärung*, Hg. Hanns-Peter Neumann (Berlin/New York 2009), 203-270.

> Bewegungsgesetze ihre vorherige Ordnung wieder erlangen. Der Mechanismus reicht also für sich allein nicht aus, um die Entstehung organisierter Körper zu erklären.[144]

Die Uhr vermag bei Ploucquet nur den Mechanismus des lebendigen Körpers zu versinnbildlichen, nicht aber wie bei La Mettrie die (Re) Organisation eines lebenden Organismus, die Ploucquet dann im Rekurs auf Leibniz' Monadentheorie streng idealistisch zu erklären versucht, indem er die Monaden zu perzeptiven, *körperbild*generierenden und -regenerierenden Lebensprinzipien macht.[145]

Auch Denis Diderot hat in seinem 1751 erschienenem *Lettre sur les sourds et les muets* auf La Mettries Uhr-Metaphorik mit der Konstruktion eines eigenen horologischen Körper-Seele-Modells Bezug genommen, ohne La Mettrie allerdings bei Namen zu nennen. Hatte La Mettrie die Seele mit dem Ticken (*timbre*), ja dann sogar mit der Hauptfeder einer Uhr verglichen, die beide *im* Horologium zu verorten seien wie die Seele *im* Körper, wird in Diderots Uhr-Metapher die Seele als Figurine *auf* der Uhr lokalisiert und somit von der Uhr selbst separiert. Diderot stellt die Seelen-Figurine als Musiker dar, der darauf „lauscht", ob sein Instrument – die Uhr – „gut gestimmt" ("bien accordé") ist. Der Musiker achtet, wenn überhaupt, nur unbewusst auf die ständigen, kaum merklichen Schwingungen der Saiten, die der verborgenen Tätigkeit des Uhrwerks entsprechen. Diderot deutet die Kontinuität dieser unmerklichen Aktivität als „unsere Existenz" ("notre existence"), die so selbstverständlich ist, dass uns deren *signa* meist

---

144) Gottfried Ploucquet, *De corporum organisatorum generatione disquisitio philosophica* (Stuttgart, 1749), § 10: "Vera quidem est propositio: ubi adest organisatio, ibi admittuntur regulae motus. Sin autem convertibilem facias propositionem hanc; falleris. Non sequitur reciproce: ubi sunt regulae motus; ibi obtinet organisatio. Leges motuum non inferunt, ut effectus ordinates modo ordinate & constanti oriatur. Attendendum enim insuper est ad coordinationem partium praeexistentem, eamque idoneam, ut talem parere queat effectum. Horologium movetur secundum regulas motus. Separa autem a se invicem rotulas, pondera &c. ac vide, num per leges motuum pristinum recuperent ordinem? Ergo mechanismus in se non sufficit explicando corporum organisatorum ortui."

145) Vgl. dazu Neumann, „Zwischen Materialismus und Idealismus," bes. S. 239-260 (Kap. 6: „Die unendliche Involution und Evolution von Bildern – monadische Repräsentationssysteme, die Erzeugung organisierter Körper und extramentale Realitäten.").

entgehen; erst das deutliche Schlagen der Glocke reizt die Seelenfigurine zur bewussten Wahrnehmung der Töne (etwa im Falle von Schmerz und Lust).[146] Das Bild der auf dem horologischen Körperautomaten

---

[146] Denis Diderot, *Lettre sur les sourds et les muets*, in ders., *Œuvres complètes*, Hg. Yvon Belaval u. a., Tôme IV (Paris, 1978), 159-160: "Monsieur, considerez l'homme automate comme une horloge ambulante; que le cœur en représente le grand ressort, et que les parties contenues dans la poitrine soient les autres pièces principales de mouvement. Imaginez dans la tête un timbre garni de petits marteaux, d'où partent une multitude infinie de fils qui se terminent à tous les points de la boîte: élevez sur ce timbre une de ces petites figures dont nous ornons le haut de nos pendules, qu'elle ait l'oreille penchée comme un musicien qui écouterait si son instrument est bien accordé; cette petite figure sera l'âme. Si plusieurs des petits cordons sont tirés dans le même instant, le timbre sera frappé de plusieurs coups, et la petite figure entendra plusieurs sons à la fois. Supposez qu'entre les cordons il y en ait certains qui soient toujours tirés; comme nous ne nous sommes assurés du bruit qui se fait le jour à Paris que par le silence de la nuit, il y aura en nous des sensations qui nous échapperont souvent par leur continuité; telle sera celle de notre existence. L'âme ne s'en aperçoit que par un retour sur elle-même, surtout dans l'état de santé. Quand on se porte bien, aucune partie du corps, ne nous instruit de son existence; si quelqu'une nous en avertit par la douleur, c'est à coup sûr que nous nous portons mal; si c'est par le plaisir, il n'est pas toujours certain que nous nous portions mieux." ("Hätte ich mit jemandem zu tun, der noch nicht die Gelegenheit gehabt hätte, abstrakte Ideen kennenzulernen, so würde ich ihm das System des menschlichen Verstandes als Skizze darbieten, und ich würde ihm sagen: Mein Herr, betrachten Sie den Menschenautomaten als eine wandelnde Uhr, so dass das Herz durch die große Feder in der Uhr repräsentiert wird und die Teile, die im Brustkörper enthalten sind, die anderen für die Bewegung wesentlichen Stücke in der Uhr sein mögen. Stellen Sie sich im Kopf eine Glocke vor, die mit kleinen Klöppeln ausgestattet ist; von dort gehen eine unendliche Menge an Fäden aus, die an allen erdenklichen Punkten des Uhrkastens enden; erheben Sie über der Glocke eine jener kleinen Figuren, mit denen wir die Spitze unserer Pendeluhren schmücken; diese Figur besitzt ein geneigtes Ohr wie ein Musiker, der darauf lauscht, ob sein Instrument gut gestimmt ist. Diese kleine Figur möge die Seele sein. Wenn nun mehrere der kleinen Schnüre im gleichen Moment gezupft würden, würde die Glocke durch mehrere Schläge geschlagen, und die kleine Figur hörte mehrere Töne auf einmal. Nehmen Sie an, dass unter den Schnüren einige wären, die ständig gezupft würden. Wie wir uns des Lärms, der tagsüber in Paris statthat, nur durch die Stille in der Nacht vergewisserten, gäbe es in uns Empfindungen, die uns durch ihre Kontinuität oft entgingen; eine solche Empfindung möge diejenige unserer Existenz sein. Die Seele würde dieser nur durch einen Rekurs auf sich selbst gewahr, vor allem im Zustand der Gesundheit. Wenn man gesund ist, belehrt uns kein Teil des Körpers von seiner Existenz; wenn irgendein Teil uns aber durch Schmerz auf sich aufmerksam macht, dann ist dies ein untrügliches Zeichen dafür, dass wir krank sind; wenn dies

thronenden Seelenfigurine stellt somit einen Rekurs auf das Leibnizsche Konzept der perzeptiven und apperzeptiven Monaden vor: Die Seelenfigurine ist der Körperuhr unmittelbar gegenwärtig, ohne doch *in* dieser verankert zu sein, wie das bei La Mettrie der Fall ist. Mit dem *homme automate* als *horloge ambulante* bezieht sich Diderot nun zwar auch auf La Mettries Uhrenmensch, wendet sich mit der Situierung der Seelenfigurine *sur le timbre* aber gleichzeitig gegen ihn. Die Seele, so scheint es wenigstens, ist bei Diderot kein integrativer Bestandteil des horologischen Körperautomaten, sondern steht deutlich über diesem. Sie verhält sich zu ihm wie der Musiker zu seinem Instrument. Zwar entzieht sich Diderot in seinem *Lettre sur les sourds et les muets* einer dezidierten Bestimmung der Seele. Dies ändert jedoch nichts daran, dass Diderot deren ornamentales, die Körpermaschine schmückendes und zugleich übersteigendes Dasein in den Vordergrund holt.

Mit der Erfindung des Uhrwerkmechanismus und der Verbreitung des Horologium im späten Mittelalter und in der Renaissance geht die metaphorische Funktionalisierung der horologischen Technologie zu diversen Beschreibungs- und Erklärungsmodellen des Verhältnisses von Gott und Welt und von Körper und Seele einher. Die Uhr- Metapher akkumuliert diverse Leib-Seele-Vorstellungen und kennzeichnet den Körper, die Seele oder die Leibseele als in sich geschlossene, selbstregulative Lebenssysteme. An ihrem Gebrauch zwischen 1450 und 1750 lassen sich – freilich aufgrund des hier bloß selektiv verwendeten Quellenmaterials nur heuristisch und konjektural – bestimmte historische Entwicklungslinien nachzeichnen. Nur langsam nämlich löste sich die Uhr-Metapher von der Versinnbildlichung der den Körper temperierenden, regulierenden, bewegenden und verlebendigenden immateriellen Kraft der Seele. Noch bei Leibniz konnte die Uhr als Gleichnis für den autarken, völlig vom Körper separierten autonomen Seelenautomaten stehen. Erst bei La Mettrie lässt sich mit der Auflösung des klassischen Seelenbegriffs ein bei Descartes schon vorgedachter, aber nicht konsequent vollzogener Transfer der Uhr-Metapher auf den ausschließlich natürlich-materiell, d. h. neurophysiologisch, organisierten

---

durch Vergnügen geschieht, dann ist es nicht immer sicher, dass es uns gesundheitlich wirklich besser geht."). – Den Hinweis auf diese Stelle verdanke ich meiner Kollegin Annette Graczyk.

Körper konstatieren. Dem idealistischen Monismus der Leibnizianer setzte La Mettrie seinen materialistischen Monismus entgegen. Beide Theorien beeinflussten den Gebrauch der Uhr-Metapher in der Aufklärung nachhaltig.

# Index

www.ingramcontent.com/pod-product-compliance
Lightning Source LLC
LaVergne TN
LVHW020652300926
849161LV00004B/920
* 9 7 8 9 0 0 4 1 9 1 8 1 5 *